Pitman Research Notes in Mathematics Series

Submission of proposals for consideration
Suggestions for publication, in the form of outlines and representative samples, are invited by the Editorial Board for assessment. Intending authors should approach one of the main editors or another member of the Editorial Board, citing the relevant AMS subject classifications. Alternatively, outlines may be sent directly to the publisher's offices. Refereeing is by members of the board and other mathematical authorities in the topic concerned, throughout the world.

Preparation of accepted manuscripts
On acceptance of a proposal, the publisher will supply full instructions for the preparation of manuscripts in a form suitable for direct photo-lithographic reproduction. Specially printed grid sheets are provided and a contribution is offered by the publisher towards the cost of typing. Word processor output, subject to the publisher's approval, is also acceptable.

Illustrations should be prepared by the authors, ready for direct reproduction without further improvement. The use of hand-drawn symbols should be avoided wherever possible, in order to maintain maximum clarity of the text.

The publisher will be pleased to give any guidance necessary during the preparation of a typescript, and will be happy to answer any queries.

Important note
In order to avoid later retyping, intending authors are strongly urged not to begin final preparation of a typescript before receiving the publisher's guidelines and special paper. In this way it is hoped to preserve the uniform appearance of the series.

Longman Scientific & Technical
Longman House
Burnt Mill
Harlow, Essex, UK
(tel (0279) 26721)

Titles in this series

1 Improperly posed boundary value problems
  A Carasso and A P Stone
2 Lie algebras generated by finite dimensional ideals
  I N Stewart
3 Bifurcation problems in nonlinear elasticity
  R W Dickey
4 Partial differential equations in the complex domain
  D L Colton
5 Quasilinear hyperbolic systems and waves
  A Jeffrey
6 Solution of boundary value problems by the method of integral operators
  D L Colton
7 Taylor expansions and catastrophes
  T Poston and I N Stewart
8 Function theoretic methods in differential equations
  R P Gilbert and R J Weinacht
9 Differential topology with a view to applications
  D R J Chillingworth
10 Characteristic classes of foliations
  H V Pittie
11 Stochastic integration and generalized martingales
  A U Kussmaul
12 Zeta-functions: An introduction to algebraic geometry
  A D Thomas
13 Explicit a priori inequalities with applications to boundary value problems
  V G Sigillito
14 Nonlinear diffusion
  W E Fitzgibbon III and H F Walker
15 Unsolved problems concerning lattice points
  J Hammer
16 Edge-colourings of graphs
  S Fiorini and R J Wilson
17 Nonlinear analysis and mechanics: Heriot-Watt Symposium Volume I
  R J Knops
18 Actions of fine abelian groups
  C Kosniowski
19 Closed graph theorems and webbed spaces
  M De Wilde
20 Singular perturbation techniques applied to integro-differential equations
  H Grabmüller
21 Retarded functional differential equations: A global point of view
  S E A Mohammed
22 Multiparameter spectral theory in Hilbert space
  B D Sleeman
24 Mathematical modelling techniques
  R Aris
25 Singular points of smooth mappings
  C G Gibson
26 Nonlinear evolution equations solvable by the spectral transform
  F Calogero

27 Nonlinear analysis and mechanics: Heriot-Watt Symposium Volume II
  R J Knops
28 Constructive functional analysis
  D S Bridges
29 Elongational flows: Aspects of the behaviour of model elasticoviscous fluids
  C J S Petrie
30 Nonlinear analysis and mechanics: Heriot-Watt Symposium Volume III
  R J Knops
31 Fractional calculus and integral transforms of generalized functions
  A C McBride
32 Complex manifold techniques in theoretical physics
  D E Lerner and P D Sommers
33 Hilbert's third problem: scissors congruence
  C-H Sah
34 Graph theory and combinatorics
  R J Wilson
35 The Tricomi equation with applications to the theory of plane transonic flow
  A R Manwell
36 Abstract differential equations
  S D Zaidman
37 Advances in twistor theory
  L P Hughston and R S Ward
38 Operator theory and functional analysis
  I Erdelyi
39 Nonlinear analysis and mechanics: Heriot-Watt Symposium Volume IV
  R J Knops
40 Singular systems of differential equations
  S L Campbell
41 N-dimensional crystallography
  R L E Schwarzenberger
42 Nonlinear partial differential equations in physical problems
  D Graffi
43 Shifts and periodicity for right invertible operators
  D Przeworska-Rolewicz
44 Rings with chain conditions
  A W Chatters and C R Hajarnavis
45 Moduli, deformations and classifications of compact complex manifolds
  D Sundararaman
46 Nonlinear problems of analysis in geometry and mechanics
  M Atteia, D Bancel and I Gumowski
47 Algorithmic methods in optimal control
  W A Gruver and E Sachs
48 Abstract Cauchy problems and functional differential equations
  F Kappel and W Schappacher
49 Sequence spaces
  W H Ruckle
50 Recent contributions to nonlinear partial differential equations
  H Berestycki and H Brezis
51 Subnormal operators
  J B Conway

52 Wave propagation in viscoelastic media
F Mainardi

53 Nonlinear partial differential equations and their applications: Collège de France Seminar. Volume I
H Brezis and J L Lions

54 Geometry of Coxeter groups
H Hiller

55 Cusps of Gauss mappings
T Banchoff, T Gaffney and C McCrory

56 An approach to algebraic K-theory
A J Berrick

57 Convex analysis and optimization
J-P Aubin and R B Vintner

58 Convex analysis with applications in the differentiation of convex functions
J R Giles

59 Weak and variational methods for moving boundary problems
C M Elliott and J R Ockendon

60 Nonlinear partial differential equations and their applications: Collège de France Seminar. Volume II
H Brezis and J L Lions

61 Singular systems of differential equations II
S L Campbell

62 Rates of convergence in the central limit theorem
Peter Hall

63 Solution of differential equations by means of one-parameter groups
J M Hill

64 Hankel operators on Hilbert space
S C Power

65 Schrödinger-type operators with continuous spectra
M S P Eastham and H Kalf

66 Recent applications of generalized inverses
S L Campbell

67 Riesz and Fredholm theory in Banach algebra
B A Barnes, G J Murphy, M R F Smyth and T T West

68 Evolution equations and their applications
F Kappel and W Schappacher

69 Generalized solutions of Hamilton-Jacobi equations
P L Lions

70 Nonlinear partial differential equations and their applications: Collège de France Seminar. Volume III
H Brezis and J L Lions

71 Spectral theory and wave operators for the Schrödinger equation
A M Berthier

72 Approximation of Hilbert space operators I
D A Herrero

73 Vector valued Nevanlinna Theory
H J W Ziegler

74 Instability, nonexistence and weighted energy methods in fluid dynamics and related theories
B Straughan

75 Local bifurcation and symmetry
A Vanderbauwhede ,

76 Clifford analysis
F Brackx, R Delanghe and F Sommen

77 Nonlinear equivalence, reduction of PDEs to ODEs and fast convergent numerical methods
E E Rosinger

78 Free boundary problems, theory and applications. Volume I
A Fasano and M Primicerio

79 Free boundary problems, theory and applications. Volume II
A Fasano and M Primicerio

80 Symplectic geometry
A Crumeyrolle and J Grifone

81 An algorithmic analysis of a communication model with retransmission of flawed messages
D M Lucantoni

82 Geometric games and their applications
W H Ruckle

83 Additive groups of rings
S Feigelstock

84 Nonlinear partial differential equations and their applications: Collège de France Seminar. Volume IV
H Brezis and J L Lions

85 Multiplicative functionals on topological algebras
T Husain

86 Hamilton-Jacobi equations in Hilbert spaces
V Barbu and G Da Prato

87 Harmonic maps with symmetry, harmonic morphisms and deformations of metrics
P Baird

88 Similarity solutions of nonlinear partial differential equations
L Dresner

89 Contributions to nonlinear partial differential equations
C Bardos, A Damlamian, J I Díaz and J Hernández

90 Banach and Hilbert spaces of vector-valued functions
J Burbea and P Masani

91 Control and observation of neutral systems
D Salamon

92 Banach bundles, Banach modules and automorphisms of C*-algebras
M J Dupré and R M Gillette

93 Nonlinear partial differential equations and their applications: Collège de France Seminar. Volume V
H Brezis and J L Lions

94 Computer algebra in applied mathematics: an introduction to MACSYMA
R H Rand

95 Advances in nonlinear waves. Volume I
L Debnath

96 FC-groups
M J Tomkinson

97 Topics in relaxation and ellipsoidal methods
M Akgül

98 Analogue of the group algebra for topological semigroups
H Dzinotyiweyi

99 Stochastic functional differential equations
S E A Mohammed

100 Optimal control of variational inequalities
V Barbu

101 Partial differential equations and dynamical systems
W E Fitzgibbon III

102 Approximation of Hilbert space operators. Volume II
C Apostol, L A Fialkow, D A Herrero and D Voiculescu

103 Nondiscrete induction and iterative processes
V Ptak and F-A Potra

104 Analytic functions – growth aspects
O P Juneja and G P Kapoor

105 Theory of Tikhonov regularization for Fredholm equations of the first kind
C W Groetsch

106 Nonlinear partial differential equations and free boundaries. Volume I
J I Díaz

107 Tight and taut immersions of manifolds
T E Cecil and P J Ryan

108 A layering method for viscous, incompressible L_p flows occupying R^n
A Douglis and E B Fabes

109 Nonlinear partial differential equations and their applications: Collège de France Seminar. Volume VI
H Brezis and J L Lions

110 Finite generalized quadrangles
S E Payne and J A Thas

111 Advances in nonlinear waves. Volume II
L Debnath

112 Topics in several complex variables
E Ramírez de Arellano and D Sundararaman

113 Differential equations, flow invariance and applications
N H Pavel

114 Geometrical combinatorics
F C Holroyd and R J Wilson

115 Generators of strongly continuous semigroups
J A van Casteren

116 Growth of algebras and Gelfand–Kirillov dimension
G R Krause and T H Lenagan

117 Theory of bases and cones
P K Kamthan and M Gupta

118 Linear groups and permutations
A R Camina and E A Whelan

119 General Wiener–Hopf factorization methods
F-O Speck

120 Free boundary problems: applications and theory, Volume III
A Bossavit, A Damlamian and M Fremond

121 Free boundary problems: applications and theory, Volume IV
A Bossavit, A Damlamian and M Fremond

122 Nonlinear partial differential equations and their applications: Collège de France Seminar. Volume VII
H Brezis and J L Lions

123 Geometric methods in operator algebras
H Araki and E G Effros

124 Infinite dimensional analysis–stochastic processes
S Albeverio

125 Ennio de Giorgi Colloquium
P Krée

126 Almost-periodic functions in abstract spaces
S Zaidman

127 Nonlinear variational problems
A Marino, L Modica, S Spagnolo and M Degiovanni

128 Second-order systems of partial differential equations in the plane
L K Hua, W Lin and C-Q Wu

129 Asymptotics of high-order ordinary differential equations
R B Paris and A D Wood

130 Stochastic differential equations
R Wu

131 Differential geometry
L A Cordero

132 Nonlinear differential equations
J K Hale and P Martinez-Amores

133 Approximation theory and applications
S P Singh

134 Near-rings and their links with groups
J D P Meldrum

135 Estimating eigenvalues with *a posteriori/a priori* inequalities
J R Kuttler and V G Sigillito

136 Regular semigroups as extensions
F J Pastijn and M Petrich

137 Representations of rank one Lie groups
D H Collingwood

138 Fractional calculus
G F Roach and A C McBride

139 Hamilton's principle in continuum mechanics
A Bedford

140 Numerical analysis
D F Griffiths and G A Watson

141 Semigroups, theory and applications. Volume I
H Brezis, M G Crandall and F Kappel

142 Distribution theorems of L-functions
D Joyner

143 Recent developments in structured continua
D De Kee and P Kaloni

144 Functional analysis and two-point differential operators
J Locker

145 Numerical methods for partial differential equations
S I Hariharan and T H Moulden

146 Completely bounded maps and dilations
V I Paulsen

147 Harmonic analysis on the Heisenberg nilpotent Lie group
W Schempp

148 Contributions to modern calculus of variations
L Cesari

149 Nonlinear parabolic equations: qualitative properties of solutions
L Boccardo and A Tesei

150 From local times to global geometry, control and physics
K D Elworthy

151 A stochastic maximum principle for optimal control of diffusions
U G Haussmann

152 Semigroups, theory and applications. Volume II
H Brezis, M G Crandall and F Kappel

153 A general theory of integration in function spaces
P Muldowney

154 Oakland Conference on partial differential equations and applied mathematics
L R Bragg and J W Dettman

155 Contributions to nonlinear partial differential equations. Volume II
J I Díaz and P L Lions

156 Semigroups of linear operators: an introduction
A C McBride

157 Ordinary and partial differential equations
B D Sleeman and R J Jarvis

158 Hyperbolic equations
F Colombini and M K V Murthy

159 Linear topologies on a ring: an overview
J S Golan

160 Dynamical systems and bifurcation theory
M I Camacho, M J Pacifico and F Takens

161 Branched coverings and algebraic functions
M Namba

162 Perturbation bounds for matrix eigenvalues
R Bhatia

163 Defect minimization in operator equations: theory and applications
R Reemtsen

164 Multidimensional Brownian excursions and potential theory
K Burdzy

165 Viscosity solutions and optimal control
R J Elliott

166 Nonlinear partial differential equations and their applications. Collège de France Seminar. Volume VIII
H Brezis and J L Lions

167 Theory and applications of inverse problems
H Haario

168 Energy stability and convection
G P Galdi and B Straughan

169 Additive groups of rings. Volume II
S Feigelstock

170 Numerical analysis 1987
D F Griffiths and G A Watson

171 Surveys of some recent results in operator theory. Volume I
J B Conway and B B Morrel

172 Amenable Banach algebras
J-P Pier

173 Pseudo-orbits of contact forms
A Bahri

174 Poisson algebras and Poisson manifolds
K H Bhaskara and K Viswanath

175 Maximum principles and eigenvalue problems in partial differential equations
P W Schaefer

176 Mathematical analysis of nonlinear, dynamic processes
K U Grusa

177 Cordes' two-parameter spectral representation theory
D F McGhee and R H Picard

178 Equivariant K-theory for proper actions
N C Phillips

179 Elliptic operators, topology and asymptotic methods
J Roe

180 Nonlinear evolution equations
J K Engelbrecht, V E Fridman and E N Pelinovski

181 Nonlinear partial differential equations and their applications. Collège de France Seminar. Volume IX
H Brezis and J L Lions

182 Critical points at infinity in some variational problems
A Bahri

183 Recent developments in hyperbolic equations
L Cattabriga, F Colombini, M K V Murthy and S Spagnolo

184 Optimization and identification of systems governed by evolution equations on Banach space
N U Ahmed

185 Free boundary problems: theory and applications. Volume I
K H Hoffmann and J Sprekels

186 Free boundary problems: theory and applications. Volume II
K H Hoffmann and J Sprekels

187 An introduction to intersection homology theory
F Kirwan

188 Derivatives, nuclei and dimensions on the frame of torsion theories
J S Golan and H Simmons

189 Theory of reproducing kernels and its applications
S Saitoh

190 Volterra integrodifferential equations in Banach spaces and applications
G Da Prato and M Iannelli

191 Nest algebras
K R Davidson

192 Surveys of some recent results in operator theory. Volume II
J B Conway and B B Morrel

193 Nonlinear variational problems. Volume II
A Marino and M K Murthy

194 Stochastic processes with multidimensional parameter
M E Dozzi

195 Prestressed bodies
D Iesan

196 Hilbert space approach to some classical transforms
R H Picard

197 Stochastic calculus in application
J R Norris

198 Radical theory
B J Gardner

199 The C* – algebras of a class of solvable Lie groups
X Wang

200 Stochastic analysis, path integration and dynamics
D Elworthy

201 Riemannian geometry and holonomy groups
S Salamon

202 Strong asymptotics for extremal errors and polynomials associated with Erdös type weights
D S Lubinsky

203 Optimal control of diffusion processes
V S Borkar

204 Rings, modules and radicals
B J Gardner

205 Numerical studies for nonlinear Schrödinger equations
B M Herbst and J A C Weideman

206 Distributions and analytic functions
R D Carmichael and D Mitrović

207 Semicontinuity, relaxation and integral representation in the calculus of variations
G Buttazzo

208 Recent advances in nonlinear elliptic and parabolic problems
P Bénilan, M Chipot, L Evans and M Pierre

209 Model completions, ring representations and the topology of the Pierce sheaf
A Carson

210 Retarded dynamical systems
G Stepan

211 Function spaces, differential operators and nonlinear analysis
L Paivarinta

212 Analytic function theory of one complex variable
C C Yang, Y Komatu and K Niino

213 Elements of stability of visco-elastic fluids
J Dunwoody

214 Jordan decompositions of generalised vector measures
K D Schmidt

215 A mathematical analysis of bending of plates with transverse shear deformation
C Constanda

216 Ordinary and partial differential equations Vol II
B D Sleeman and R J Jarvis

217 Hilbert modules over function algebras
R G Douglas and V I Paulsen

218 Graph colourings
R Wilson and R Nelson

219 Hardy-type inequalities
A Kufner and B Opic

220 Nonlinear partial differential equations and their applications. College de France Seminar Volume X
H Brezis and J L Lions

221 Workshop on dynamical systems
E Shiels and Z Coelho

222 Geometry and analysis in nonlinear dynamics
H W Broer and F Takens

223 Fluid dynamical aspects of combustion theory
M Onofri and A Tesei

224 Approximation of Hilbert space operators. Volume I. 2nd edition
D Herrero

225 Operator Theory: Proceedings of the 1988 GPOTS–Wabash conference
J B Conway and B B Morrel

226 Local cohomology and localization
J L Bueso Montero, B Torrecillas Jover and A Verschoren

227 Sobolev spaces of holomorphic functions
F Beatrous and J Burbea

228 Numerical analysis. Volume III
D F Griffiths and G A Watson

229 Recent developments in structured continua. Volume III
D De Kee and P Kaloni

230 Boolean methods in interpolation and approximation
F J Delvos and W Schempp

231 Further advances in twistor theory, Volume 1
L J Mason and L P Hughston

232 Further advances in twistor theory, Volume 2
L J Mason and L P Hughston

233 Geometry in the neighborhood of invariant manifolds of maps and flows and linearization
U Kirchgraber and K Palmer

234 Quantales and their applications
K I Rosenthal

Operator Theory: Proceedings of the 1988 GPOTS- Wabash conference

J. B. Conway & B. B. Morrel

Indiana University/Indiana University

Purdue University at Indianapolis

Operator Theory: Proceedings of the 1988 GPOTS-Wabash conference

Longman
Scientific &
Technical

Longman Scientific & Technical,
Longman Group UK Limited,
Longman House, Burnt Mill, Harlow
Essex CM20 2JE, England
and Associated Companies throughout the world.

Copublished in the United States with
John Wiley & Sons, Inc., 605 Third Avenue, New York, NY 10158

© Longman Group UK Limited 1990

First published 1990

AMS classification: 46L05, 46L40, 46L80, 19K56, 47D25

ISSN 0269-3674

British Library Cataloguing in Publication Data
Operator theory
  1. Mathematics, Operators
  I. Conway, John B.   II. Morrel, Bernard B. (Bernard Baldwin), *1940–*   III. Series
  515.7′24

ISBN 0-582-06190-3

Library of Congress Cataloging-in-Publication Data
Operator theory/John B. Conway & Bernard B. Morrel (editors).
    p.   cm. — (Pitman research notes in mathematics series ; 225)
    "Proceedings of the Gpots–Wabash Conference of May, 1988"—CIP galley.
    ISBN 0-470-21594-1
    1. Operator theory—Congresses. I. Conway, John B. II. Morrel, Bernard B. III. Series.
QA329.062   1990   89-38852
515′.724—dc20   CIP

Printed and bound in Great Britain
by Biddles Ltd, Guildford and King's Lynn

TABLE OF CONTENTS

PREFACE

Quantizing the Fredholm index

    W. Arveson 1

Integral representations of some Hankel operators on the Hardy and
Bergman spaces

    P. G. Ghatage 33

Full group C*-algebras and the generalized Kadison conjecture

    R. Ji and S. Pedersen 43

The corona construction

    Gert K. Pedersen 49

Crossed product C*-algebras by compact group actions vs. fixed point algebras

    C. Peligrad 93

Shifts on certain operator algebras

    G. L. Price 99

Measure-theoretic properties of the central topology on the set of the factorial
states of a C*-algebra. Central reduction

    S. Teleman 107

Preface

This volume is the proceedings of the combined meeting of GPOTS (Great Plains Operator Theory Seminar) and the Wabash Extramural Functional Analysis Seminar which was held at IUPUI (Indiana University—Purdue University at Indianapolis) from May 7 to May 10, 1988. The featured speakers at the conference were Bill Arveson, Ron Douglas, and Gert K. Pedersen, each of whom gave three superb one-hour lectures. All of these lectures, as well as the shorter talks which were contributed by the participants, were truly outstanding. The talks were so good, in fact, that, as the meeting progressed, several of the participants suggested that the proceedings be published. This volume is the result of their suggestions.

There are always many people who contribute to the success of an enterprise of this magnitude, and the 1988 GPOTS—Wabash is no exception. Financial support for the conference came from a wide variety of sources. Thanks first to the National Science Foundation for their continuing support of GPOTS. Thanks to Carol Nathan, Associate Dean of the Faculties at IUPUI, whose funds enabled us to rent the lecture rooms at the University Conference Center. Thanks also to Dr. Wendell McBurney, Dean of Research and Sponsored Programs at IUPUI, to the Mathematics Departments of Indiana University, IUPUI, and Purdue University, and to IMA (Institute for Mathematics and Its Applications) for their financial support of the conference. Last, but certainly not least, thanks to the Wabash Extramural Functional Analysis Seminar for its joint participation and its financial support.

Special thanks are due to Mr. John D. Short, Director of the University Conference Center and to Mike Prakel and the entire staff of the Conference Center. Their service, assistance, and friendly demeanor matched the quality of the facilities, which are truly world-class. Thanks also to Linda Eldridge and the staff of the University Place Hotel (formerly, the Lincoln Hotel) for their excellent service and to Garry Owens and Maxine Petri of the IUPUI Mathematics Department, who handled many of the details of the conference expertly and efficiently. Thanks also to Frank Gilfeather and Norberto Salinas, whose encouragement, support, and valuable suggestions made the conference a much better one than it otherwise would have been.

Finally, thanks to all of the speakers and participants in the conference, to all of the contributors to this volume, and to Jan Want of Texas A&M University, who did a superb job of preparing the typescript.

John B. Conway

Bernard B. Morrel

W. Arveson

Quantizing the Fredholm index

§0. Introduction

This paper is a somewhat expanded version of a series of lectures given by the author at the Great Plains Operator Theory Seminar held in Indianapolis in May, 1988. The subject matter is index theory for semigroups of *-endomorphisms of $\mathcal{B}(H)$ (E_0-semigroups), the study of which was initiated by Powers in [7], [8], Powers and Robinson in [9], and continued by the author in [2], [3].

Our objective is to give a painless exposition of the theory of continuous tensor product systems, their role in the classification of E_0-semigroups [2], [3], and to make the point that this theory is a quantized version of the Fredholm index theory of unbounded symmetric operators. In keeping with the first of these two goals, we have taken most of the proofs out and put most of the intuition in. Where there is some novelty in the exposition, as in §3, we have included essentially complete arguments whenever it seemed necessary and appropriate to do so.

The notation is standard and requires no elaboration here. But the reader should keep in mind that our methods require all Hilbert spaces to be *separable*.

§1. Fredholm Index of a Semigroup of Isometries

Let $U = \{U_t : t \geq 0\}$ be a semigroup of isometries acting on a Hilbert space H. The generator A of U, appropriately defined as

$$A = \frac{d}{dt} U_t|_{t=0},$$

is a densely defined unbounded skew-symmetric operator on H for which

$$U_t = e^{tA}, \quad t \geq 0.$$

Recall that any skew-symmetric operator A gives rise to two deficiency subspaces $\mathcal{E}_+, \mathcal{E}_-$ of its domain, defined by

$$\mathcal{E}_\pm = \{\xi \in \mathcal{D}_A : A\xi = \pm\xi\}.$$

If one of $\mathcal{E}_\pm$ is finite dimensional then the *index* of A can be defined as

$$\text{index } A = \dim \mathcal{E}_+ - \dim \mathcal{E}_-.$$

In the case above where A is the generator of a semigroup of isometries one has $\mathcal{E}_- = \{0\}$, and hence index A is always defined and non-negative.

This index has the following stability. If B is any *bounded* skew-adjoint operator on H, then $A + B$ is also the generator of a semigroup of isometries and one has

$$\text{index } (A + B) = \text{ index } A.$$

Let us first reformulate this definition so as to get rid of unbounded operators. If $U = \{U_t : t \geq 0\}$ is a semigroup of isometries, then the projections

$$P_t = 1 - U_t U_t^*, \quad 0 \leq t < \infty$$

are increasing and we may define $P_\infty = \lim_{t \to \infty} P_t$. Clearly $0 \leq P_\infty \leq 1$, and all possibilities can occur. If $P_\infty = 0$ (equivalently, if each U_t is a unitary operator) we define index $U = 0$. If $P_\infty \neq 0$, then there is a unique spectral measure Q, defined on $[0, \infty)$ and taking values in $\mathcal{B}(P_\infty H)$, which satisfies

$$Q([0, t]) = P_t|_{P_\infty H}, \quad t \geq 0.$$

It is not hard to see that the measure class of Q is Lebesgue measure and that Q has uniform multiplicity $n = 1, 2, \ldots, \infty$. We define the index of U to be that integer n.

This definition agrees with the preceding one given in terms of generators. It is also possible to reformulate (and generalize) the above stability property in a form appropriate for semigroups of isometries, but we will not have to do that here.

There are two natural operations that one can perform on pairs of such semigroups U, V, namely

$$\textit{direct sum} : (U \oplus V)_t = U_t \oplus V_t, \ t \geq 0$$

$$\textit{tensor product} : (U \otimes V)_t = U_t \otimes V_t, \ t \geq 0.$$

Consider first the direct sum. If we set $W_t = U_t \oplus V_t, \ t \geq 0$, then

$$1 - W_t W_t^* = (1 - U_t U_t^*) \oplus (1 - V_t V_t^*)$$

and hence the spectral measures for U, V, W are related by

$$Q_W = Q_U \oplus Q_V.$$

Since Q_U and Q_V are mutually absolutely continuous their multiplicities add in the obvious way, and we conclude that

$$\text{index}(U \oplus V) = \text{index } U + \text{index } V.$$

On the other hand, the arithmetic of the index of tensor products breaks down completely: the index of $U \otimes V$ is infinite in all cases where index $U > 0$ and index $V > 0$. Thus we shall make use of only the direct sum operation when discussing the Fredholm index of such semigroups.

§2. E_0-semigroups

Let $\alpha = \{\alpha_t : t \geq 0\}$ be a semigroup of normal $*$-homomorphisms of $\mathcal{B}(H)$ into itself satisfying $\alpha_t(1) = 1$ for all $t \geq 0$, and which is continuous in the sense that $\langle \alpha_t(A)\xi, \eta \rangle$ is a continuous function of t for fixed $A \in \mathcal{B}(H)$ and fixed $\xi, \eta \in H$. Following Powers [7], [8] we shall call such an α and E_0-semigroup.

Two E_0-semigroups α, β are said to be *conjugate* if there is a $*$-isomorphism $\theta :$ $\mathcal{B}(H_\alpha) \to \mathcal{B}(H_\beta)$ such that

$$\beta_t(B) = \theta\alpha_t(\theta^{-1}(B)), \quad t \geq 0, \ B \in \mathcal{B}(H_\beta).$$

θ is necessarily normal, and in fact is implemented by a unitary operator $W : H_\alpha \to H_\beta$ in the usual way

$$\theta(A) = WAW^*.$$

Thus, all properties of E_0-semigroups are preserved under conjugacy.

Every E_0-semigroup α has a generator δ, which is a densely defined $*$-derivation of $\mathcal{B}(H)$. $\delta(A)$ is defined as the limit in the strong operator topology

$$\delta(A) = \lim_{t \to 0} t^{-1}(\alpha_t(A) - A),$$

and of course the domain $\mathcal{D}$ of δ is the set of all $A \in \mathcal{B}(H)$ for which this limit exists. If B is a bounded self-adjoint operator on H, then

(2.1) $$\delta'(A) = \delta(A) + i(BA - AB)$$

defines another derivation on $\mathcal{D}$, and it is a fact that there is a unique E_0-semigroup α' having generator δ'. We will say (somewhat imprecisely) that δ' is a perturbation of δ by a bounded operator.

3

This definition catches the idea of the perturbation theory that we need to consider, but it is hard to manage in practice because different generators may well have domains with nothing in common but 0.

We have to reformulate and generalize this relation between δ and $\dot{\delta}'$ into a more tractable form. Let α, β be two E_0-semigroups which *act on the same von Neumann algebra* $\mathcal{B}(H)$. α and β are said to be *exterior equivalent* (written $\alpha \underset{e}{\sim} \beta$) if there is a strongly continuous family of unitary operators $\{U_t : t \geq 0\}$ in $\mathcal{B}(H)$ satisfying

(2.2)
$$\text{(i)} \quad U_{s+t} = U_s \alpha_s(U_t), \quad s, t \geq 0$$
$$\text{(ii)} \quad \beta_t(A) = U_t \alpha_t(A) U_t^*, \quad A \in \mathcal{B}(H), \ t \geq 0.$$

The cocycle condition (i) guarantees that $\{U_t \alpha_t U_t^* : t \geq 0\}$ does indeed define an E_0-semigroup. Note too that (i) implies $U_0 = U_0^2$, and hence $U_0 = 1$.

In order to see that exterior equivalence generalizes the above perturbation theory, suppose α, α' are two E_0-semigroups whose generators δ, δ' are related as in (2.1) with a bounded self-adjoint operator B. By a classical iteration technique, one can show that the differential equation

$$\frac{d}{dt} U(t) = iU(t)\alpha_t(B), \quad t \geq 0$$
$$U(0) = 1$$

has a unique solution on $0 \leq t < \infty$ and that the solution $\{U(t) : t \geq 0\}$ satisfies (2.2) (for more detail, see [11, p. 282]). Indeed, since B is bounded the cocycle $U(\cdot)$ is *norm-continuous*. This will not always be the case for the more general cocycles appearing in (2.2). Nevertheless, it is useful to think of the exterior equivalence relation as a precise formulation of the formal idea of perturbing the generator of α by an "unbounded operator" to obtain the generator of α'.

Definition 2.3. *Two E_0-semigroups α, β are said to be outer conjugate if β is conjugate to an E_0-semigroup β' acting on $\mathcal{B}(H_\alpha)$ with the property that $\beta' \underset{e}{\sim} \alpha$.*

The reader will recognize that this is a variation of Connes' definition of outer conjugacy of automorphism groups [4], [10] into a form appropriate for E_0-semigroups. The intuition is this; if conjugacy of E_0-semigroups is analogous to unitary equivalence of operators, then outer conjugacy is analogous to the unitary equivalence of operators modulo compacts. We will show that these are more than analogies in §3.

We conclude this section with a few general comments about the structure of E_0-semigroups. Such a semigroup $\alpha = \{\alpha_t : t \geq 0\}$ may of course have the property that

$$\alpha_t(\mathcal{B}(H)) = \mathcal{B}(H)$$

for every $t \geq 0$. An E_0-semigroup with this property is called *degenerate*. These semi-groups were excluded from consideration in [2]; however, it is actually more convenient to include them. If α is degenerate then each α_t is an *automorphism* of $\mathcal{B}(H)$, and it is not hard to show that there is a strongly continuous unitary group $\{V_t : t \in \mathbb{R}\}$ which implements α in the sense that

$$\alpha_t(A) = V_t A V_t^*, \quad A \in \mathcal{B}(H), \ t \geq 0.$$

If β is another degenerate E_0-semigroup acting on $\mathcal{B}(H)$ and $\{W_t : t \in \mathbb{R}\}$ is its corresponding implementing unitary group, then the family of unitary operators

$$U_t = W_t V_t^*, \quad t \geq 0$$

is strongly continuous and satisfies $U_{s+t} = U_s \alpha_s(U_t)$ and $\beta_t = U_t \alpha_t U_t^*$ for all $s, t \geq 0$. We conclude that *all degenerate E_0-semigroups are outer conjugate*. This is analogous to the fact that two normal operators having the same essential spectrum are unitarily equivalent modulo compacts. In particular, outer conjugacy is plainly a far weaker relation than conjugacy.

Now let α be a nondegenerate E_0-semigroup. Fix $t > 0$ and consider the single endomorphism $\alpha_t : \mathcal{B}(H) \to \mathcal{B}(H)$. It is rather easy to show that there is a (necessarily infinite) sequence $V_1, V_2, \ldots$ of isometries in $\mathcal{B}(H)$, having mutually orthogonal ranges, which implements α_t in the sense that

$$\alpha_t(A) = \sum_{n=1}^{\infty} V_n A V_n^*, \quad A \in \mathcal{B}(H).$$

It follows that $\sum_n V_n V_n^* = 1$. The sequence $\{V_1, V_2, \ldots\}$ is certainly not unique, but one can show that if $\{V_1', V_2', \ldots\}$ is another such sequence, then $\{V_n\}$ and $\{V_n'\}$ generate the same norm-closed linear subspace E_t of $\mathcal{B}(H)$. As t varies over $(0, \infty)$, these operator spaces E_t vary too. In §5 we will introduce an effective way of describing these spaces, and we will discuss their precise relationship to α.

§3. Operations and Examples

Until now, we have seen no examples of nondegenerate E_0-semigroups. Here is a general procedure that will give concrete examples in certain cases. Let $U = \{U_t : t \in \mathbb{R}\}$ be a one-parameter unitary group acting on H and let γ be its associated one-parameter automorphism group: $\gamma_t(A) = U_t A U_t^*$. If M is a type I_∞ subfactor of $\mathcal{B}(H)$ which is *invariant* in the sense that $\gamma_t(M) \subseteq M$ for $t \geq 0$, then

$$\alpha_t = \gamma_t|_M, \quad t \geq 0$$

defines an E_0-semigroup acting on M. α will be nondegenerate whenever $\gamma_t(M)$ is contained properly in M for some (and therefore every) positive t.

On the other hand, if one starts with a concrete unitary group $\{U_t\}$, such as the group of translations acting in $L^2(\mathbb{R})$, it is not at all clear how to exhibit a nontrivial invariant type I_∞ subfactor M of $\mathcal{B}(H)$. Hence the construction of examples involves a somewhat more careful analysis.

Rather than simply reiterate Powers' examples (the CAR flows [7]), we present an alternate construction in order to make certain points about the more formal aspects of the relationship between the theory of E_0-semigroups and the theory of semigroups of isometries. In particular, we will show that the former is a quantized version of the latter. The E_0-semigroups one obtains by this construction will be called CCR *flows*.

We have seen that of the two operations $U \oplus V$ and $U \otimes V$ that can be performed on semigroups of isometries U and V, only the direct sum behaves well with respect to the Fredholm index. Let us now examine the situation with respect to E_0-semigroups α and β. Notice first that there is a unique E_0-semigroup $\gamma = \{\gamma_t : t \geq 0\}$ acting on $\mathcal{B}(H_\alpha \otimes H_\beta)$ which satisfies

$$\gamma_t(A \otimes B) = \alpha_t(A) \otimes \beta_t(B)$$

for all $A \in \mathcal{B}(H_\alpha)$, $B \in \mathcal{B}(H_\beta)$, $t \geq 0$. Naturally, we will denote γ by $\alpha \otimes \beta$.

On the other hand, we claim that there is no direct sum operation for E_0-semigroups. To see why, fix α and β. By replacing β with a conjugate semigroup if necessary, we can assume that α and β act on the same $\mathcal{B}(H)$. We would like to define an E_0-semigroup $\gamma = \{\gamma_t : t \geq 0\}$ acting on $\mathcal{B}(H \oplus H)$ with at *least* the following property:

$$\gamma_t \begin{pmatrix} A & 0 \\ 0 & B \end{pmatrix} = \begin{pmatrix} \alpha_t(A) & 0 \\ 0 & \beta_t(B) \end{pmatrix}, \quad t \geq 0,$$

whenever $A, B \in \mathcal{B}(H)$. However, it is not hard to see that such a semigroup γ exists if, and only if, α and β are *exterior equivalent* in the sense of (2.2). The proof is a variation of Connes' elegant trick for one-parameter automorphism groups, and is practically a verbatim restatement of the proof of Lemma 8.11.2 in [10]. In particular, it is impossible to form the "direct sum" of two E_0-semigroups which are not already outer conjugate. One may summarize this state of affairs as follows: while the only appropriate operation for semigroups of isometries is the direct sum, the only *possible* operation for E_0-semigroups is the tensor product.

These remarks suggest that the theory of E_0-semigroups should be in some sense an "exponentiated" version of the theory of semigroups of isometries. The purpose of this section is to describe the precise sense in which this is true. We present a construction whereby, starting with a semigroup U of isometries, one manufactures an E_0-semigroup α^U. The construction defines a functor from the category of semigroups of isometries to the category of E_0-semigroups. This functor combines both forms of formal quantization: the object map is first quantization for bosons and the arrow map is second quantization. Moreover, it is like exponentiation in that direct sums map to tensor products.

In more detail, let $\mathcal{S}$ be the category whose objects are semigroups of isometries and whose maps are intertwining unitary operators. Thus, $\hom(U, V)$ consists of all unitary operators $W : H_U \to H_V$ satisfying

$$WU_t = V_t W, \quad \text{for all } t \geq 0.$$

Composition of arrows has the obvious meaning. This category is a groupoid, and it admits a direct sum operation which we have already discussed in §1. For example, if $W^j \in \hom(U^j, V^j)$ for $j = 1, 2$, then $W^1 \oplus W^2$ is defined as the direct sum of unitary operators

$$W^1 \oplus W^2 : H_{U^1 \oplus U^2} = H_{U^1} \oplus H_{U^2} \to H_{V^1 \oplus V^2} = H_{V^1} \oplus H_{V^2}.$$

Let $\mathcal{E}$ be the category whose objects are E_0-semigroups and whose maps are conjugacies, so that an element of $\hom(\alpha, \beta)$ is a $*$-isomorphism $\theta : \mathcal{B}(H_\alpha) \to \mathcal{B}(H_\beta)$ which satisfies

$$\theta \circ \alpha_t = \beta_t \circ \theta, \quad t \geq 0.$$

The natural operation in this category is the tensor product described above.

In order to define a functor from $\mathcal{S}$ to $\mathcal{E}$, fix a semigroup of isometries $U = \{U_t : t \geq 0\}$ and let $H = H_U$ be the space on which U acts. We will construct a new Hilbert space e^H and an E_0-semigroup α^U which acts on $\mathcal{B}(e^H)$.

e^H is the symmetric Fock space over H, i.e., the direct sum

$$e^H = \sum_{n=0}^{\infty} H^{(n)},$$

where $H^{(n)}$ denotes the symmetric tensor product of n copies of H for $n \geq 1$ and $H^{(0)} = \mathbb{C}$. For every vector $\xi \in H$ we have a vector $\exp \xi \in e^H$ by way of

$$\exp \xi = \sum_{n=0}^{\infty} n^{-1/2} \xi^{(n)},$$

$\xi^{(n)}$ denoting $\xi \otimes \xi \otimes \ldots \otimes \xi$ (n-times) if $n \geq 1$, and $\xi^{(0)} = 1$. The relation between the inner products in H and e^H is expressed by the formula

$$\langle \exp \xi, \exp \eta \rangle = e^{\langle \xi, \eta \rangle}.$$

The vectors $\{\exp \xi : \xi \in H\}$ span e^H (see [6] for more detail). The unit vector $v = \exp(0)$ is called the *vacuum*.

Using these facts about the vectors $\{\exp \xi : \xi \in H\}$ it is apparent that every isometry $W : H \to H'$ induces a unique isometry $\tilde{W} : e^H \to e^{H'}$ having the property

$$\tilde{W}(\exp \xi) = \exp W\xi, \quad \xi \in H.$$

Moreover, by checking inner products in the same way, one can easily prove the following

Proposition 3.1. *For every vector $\xi \in H$ there is a unique unitary operator $W(\xi)$ on e^H satisfying*

$$W(\xi) : \exp \eta \mapsto e^{-\frac{1}{2}\|\xi\|^2 - \langle \eta, \xi \rangle} \exp(\eta + \xi),$$

for all $\eta \in H$.

W is a strongly continuous function from H to the unitary operators on e^H, and it satisfies the canonical commutation relations in their exponential form due to Weyl:

$$(3.2) \qquad W(\xi)W(\eta) = e^{i \, \mathrm{Im}\langle \xi, \eta \rangle} W(\xi + \eta).$$

The vacuum expectation values of W are given by

$$(3.3) \qquad \langle W(\xi)v, v \rangle = e^{-\frac{1}{2}\|\xi\|^2},$$

and vectors $\{W(\xi)v : \xi \in H\}$ span e^H.

Notice that the construction of W is natural in the sense that if $V : H \to H'$ is any unitary operator then we have

$$(3.4) \qquad \tilde{V}W_H(\xi)\tilde{V}^* = W_{H'}(V\xi), \quad \xi \in H.$$

Returning now to the semigroup $U = \{U_t : t \geq 0\} \subseteq \mathcal{B}(H)$, we obtain an E_0-semigroup acting on $\mathcal{B}(e^H)$ in the following way.

8

Proposition 3.5. *The family of operators $\{W(\xi) : \xi \in H\}$ is irreducible. Moreover, for every semigroup U of isometries acting on H there is a unique E_0-semigroup $\alpha = \{\alpha_t : t \geq 0\}$ acting on $\mathcal{B}(e^H)$ such that*

$$\alpha_t(W(\xi)) = W(U_t\xi),$$

for all $\xi \in H$, $t \geq 0$.

Proof: It is well-known that W is an irreducible Weyl system [6], [1]. It follows that each α_t is uniquely defined by the above formula, provided it exists.

For existence, we may find a one-parameter unitary group $\{V_t : t \in \mathbb{R}\}$ acting on a Hilbert space $H' \supseteq H$ such that

$$U_t = V_t|_H, \quad t \geq 0.$$

If we let $W' : H' \to \mathcal{B}(e^{H'})$ be the Weyl system associated with H' and let $\{\tilde{V}_t : t \in \mathbb{R}\}$ be the unitary group defined on $e^{H'}$ by

$$\tilde{V}_t(\exp \xi') = \exp(V_t\xi'), \quad \xi \in H', \ t \in \mathbb{R},$$

then we have

$$\tilde{V}_t W'(\xi') = W'(V_t\xi')\tilde{V}_t.$$

Hence $\beta_t(A) = \tilde{V}_t A \tilde{V}_t^*$ is a one-parameter group of $*$-automorphisms of $\mathcal{B}(e^{H'})$ satisfying

$$\beta_t(W'(\xi')) = W'(V_t\xi'), \quad t \in \mathbb{R}.$$

Let M be the von Neumann subalgebra of $\mathcal{B}(e^{H'})$ generated by $\{W'(\xi) : \xi \in H\}$. We have $\beta_t(M) \subseteq M$ for $t \geq 0$, so that $\{\beta_t : t \geq 0\}$ restricts to an E_0-semigroup acting on the von Neumann algebra M.

The inclusion $H \subseteq H'$ induces an isometry $J : e^H \to e^{H'}$ whose range is invariant under M. One checks that the map $\theta : M \to \mathcal{B}(e^H)$ defined by

$$\theta(A) = J^*AJ$$

is a $*$-isomorphism, and that $\alpha_t = \theta\beta_t\theta^{-1}$ defines the required E_0-semigroup acting on $\mathcal{B}(e^H)$. $\qquad\square$

For every $U \in \mathcal{S}$ we have constructed an E_0-semigroup of α^U acting on $\mathcal{B}(e^H)$. We show next how arrows in $\mathcal{S}$ map to arrows in $\mathcal{E}$. Let U, U' be two semigroups of isometries acting on H, H' and let $V : H \to H'$ be a unitary operator satisfying

$$V U_t = U'_t V, \quad t \geq 0.$$

We have to define a $*$-isomorphism $\Gamma(V) : \mathcal{B}(e^H) \to \mathcal{B}(e^{H'})$ with the property

(3.6) $$\Gamma(V) \circ \alpha^U = \alpha^{U'} \circ \Gamma(V).$$

This is easy: let $\tilde{V} : e^H \to e^{H'}$ be the unitary operator defined by

$$\tilde{V}(\exp \xi) = \exp(V\xi), \quad \xi \in H,$$

and put $\Gamma(V)(A) = \tilde{V} A \tilde{V}^*$, $A \in \mathcal{B}(e^H)$. From (3.4) we see that

$$\Gamma(V)(W(\xi)) = W'(V\xi), \quad \xi \in H,$$

which implies that for $t \geq 0$,

$$\Gamma(V)(\alpha_t(A)) = \alpha'_t(\Gamma(V)(A))$$

holds for all operators of the form $A = W(\xi)$.

Now let U, U' be a pair of semigroups from $\mathcal{S}$. We claim that $\alpha^{U \oplus U'}$ is naturally isomorphic to $\alpha^U \otimes \alpha^{U'}$. More precisely, for every such pair (U, U') we will exhibit an isomorphism

$$\theta(U, U') \in \text{hom}(\alpha^{U \oplus U'}, \alpha^U \otimes \alpha^{U'})$$

which is natural in the sense that if (V, V') is another pair in $\mathcal{S}$ and we choose arbitrary arrows $W \in \text{hom}(U, V)$, $W' \in \text{hom}(U', V')$, then we have a commutative diagram

(3.2)
$$
\begin{array}{ccc}
\alpha^{U \oplus U'} & \xrightarrow{\theta(U,U')} & \alpha^U \otimes \alpha^{U'} \\
{\scriptstyle \Gamma(W \oplus W')} \downarrow & & \downarrow {\scriptstyle \Gamma(W) \otimes \Gamma(W')V} \\
\alpha^{V \oplus V'} & \xrightarrow[\theta(V,V')]{} & \alpha^V \otimes \alpha^{V'}.
\end{array}
$$

Let H, H' be the respective Hilbert spaces of U, U'. If $\xi, \eta \in H$, $\xi', \eta' \in H'$, then we have

$$\langle \exp(\xi \oplus \eta), \exp(\xi' \oplus \eta') \rangle = e^{\langle \xi \oplus \eta, \xi' \oplus \eta' \rangle} =$$

$$= e^{\langle \xi, \xi' \rangle + \langle \eta, \eta' \rangle} =$$

$$= \langle \exp \xi \otimes \exp \eta, \exp \xi' \otimes \exp \eta' \rangle.$$

Hence there is a unique unitary operator $W = W_{(U,V)} : e^{H \oplus H'} \rightarrow e^H \otimes e^{H'}$ satisfying

$$W(\exp(\xi \oplus \eta)) = \exp \xi \otimes \exp \eta.$$

$\theta(U, U')(A) = WAW^*$ defines a $*$-isomorphism of $\mathcal{B}(e^{H \oplus H'})$ onto $\mathcal{B}(e^H \otimes e^{H'})$. One verifies that

$$\theta(U, U')(\alpha_t^{U \oplus U'}(A)) = (\alpha_t^U \otimes \alpha_t^{U'})(\theta(U, U')(A))$$

holds for all $A \in \mathcal{B}(e^{H \oplus H'})$ by checking it directly on operators of the form $A = W_{H \oplus H'}(\xi \oplus \xi')$, $\xi \in H$, $\xi' \in H'$, as before. Thus $\theta(U, U')$ belongs to $\hom(\alpha^{U \oplus U'}, \alpha^U \otimes \alpha^{U'})$. It is a routine matter to verify that all diagrams of the form (3.7) commute.

To summarize, we have exhibited a functor $U \rightarrow \alpha^U$ from the category $\mathcal{S}$ of semigroups of isometries to the category $\mathcal{E}$ of E_0-semigroups with the property that $\alpha^{U \oplus U'}$ is naturally isomorphic to $\alpha^U \otimes \alpha^{U'}$.

Let us now see which E_0-semigroups can be obtained from this construction. Let K be a Hilbert space of dimension n, $1 \leq n \leq +\infty$, and let $U = \{U_t : t \geq 0\}$ be the semigroup of isometries acting on the Hilbert space $L^2((0, \infty); K)$ of all K-valued L^2 functions on the positive real axis by

$$U_t f(x) = \begin{cases} f(x - t), & x > t \\ 0, & 0 < x \leq t \, . \end{cases}$$

The integer n is a unitary invariant of the semigroup U, indeed $n = \operatorname{index} U$, and of course we call U the *shift of multiplicity n*. The E_0-semigroup α^U is called the *CCR flow of rank n*. It is not clear at this point if the integer n is an invariant of α^U, but we will see in §8 that it is.

More generally, consider the case of an arbitrary semigroup U in $\mathcal{S}$. By the Wold decomposition, U can be expressed uniquely as a direct sum

(3.8) $$U_t = V_t \oplus W_t, \quad t \geq 0$$

where W is a semigroup of unitary operators and V is a semigroup of isometries which is *pure* in the sense that

$$\bigcap_{t>0} \operatorname{ran} V_t = \{0\}.$$

Of course, either summand V or W may be absent. In case V is present, it can be shown to be unitarily equivalent to the shift of multiplicity n where n is the integer $n = \operatorname{index} U = \operatorname{index} V$.

Hence α^U decomposes into a tensor product

$$\alpha^U = \alpha^V \otimes \alpha^W,$$

where α^V is conjugate to the CCR flow of rank n and α^W is a degenerate E_0-semigroup. We will see later on that when V is present, $\alpha^V \otimes \alpha^W$ is outer conjugate to α^V (c.f. corollary of Theorem 6.4). We conclude that *every E_0-semigroup which is obtained by way of the construction $U \in \mathcal{S} \to \alpha^U \in \mathcal{E}$ is either degenerate or is outer conjugate to some CCR flow, and that every CCR flow arises in this way.*

In [7], Powers gave a different construction of E_0-semigroups which utilizes the canonical anticommutation relations. If one starts with the shift of multiplicity n then one obtains the CAR flow of rank n. Powers' construction can be regarded as the analogue (for fermions) of what we have done above (for bosons). Instead of the symmetric Fock space e^H, one makes use of the antisymmetric Fock space

$$\mathcal{F}_a = \sum_{n=0}^{\infty} H^{\wedge n},$$

where $H^{\wedge n}$ is the antisymmetric tensor product of n copies of H if $n \geq 1$ and is $\mathbb{C}$ if $n = 0$. However, one does not obtain anything new from this construction because Powers has shown that the CAR flow of rank n is conjugate to the CCR flow of rank n [7], [8], [9].

§4. Numerical Index of E_0-semigroups

In [7], Powers introduced a numerical index for E_0-semigroups which takes values in $\{1, 2, \ldots, \infty\}$. This number was shown to have the "correct value" in the sense that the index of the CAR flow of rank n is n, and it is stable under bounded perturbations of the generator. However, because of certain ambiguities, it was not clear if this number was well-defined. In order to get around this difficulty, Powers and Robinson took a new approach in [9]. They defined an equivalence relation in the class of E_0-semigroups, whose classes played the role of the "index", and they showed that this more abstract index differentiates between CAR flows as well as other kinds of E_0-semigroups.

The purpose of this section is to introduce a new numerical index for E_0-semigroups. Specifically, we will construct a Hilbert space using certain structures derived from a given semigroup α; the index of α is the dimension of this Hilbert space. We will relate

12

this invariant to the Powers–Robinson index in §5 and §6, and in §8 we indicate how one calculates it.

Let α be an E_0-semigroup acting on $\mathcal{B}(H)$. By a *unit* of α we mean a strongly continuous semigroup $U = \{U_t : t \geq 0\}$ of operators in $\mathcal{B}(H)$ which satisfies $U_0 = 1$ and

(4.1) $$\alpha_t(A)U_t = U_t A, \quad t \geq 0, \ A \in \mathcal{B}(H).$$

The set of all units of α will be denoted $\mathcal{U}_\alpha$.

It follows from the work of Powers [8] that there exist E_0-semigroups α which have no units whatsoever. But if $\mathcal{U}_\alpha$ is not void, then we can define a function $c : \mathcal{U}_\alpha \times \mathcal{U}_\alpha \to \mathbb{C}$ as follows.

Proposition 4.2. *For every pair U, V in $\mathcal{U}_\alpha$, there is a unique complex number $c(U, V)$ satisfying*

$$V_t^* U_t = e^{tc(U,V)} 1, \quad t \geq 0.$$

For fixed t, the function $k_t(U, V) = e^{tc(U,V)}$ is positive definite.

Proof: The argument is very simple. Holding U and V fixed, one observes that by the commutation Formula (4.1), $V_t^* U_t$ commutes with $\mathcal{B}(H)$ for every $t \geq 0$. This gives us a continuous function $b : [0, \infty) \to \mathbb{C}$ such that $V_t^* U_t = b(t)1$. One now verifies that, from the semigroup property of U and V one has $b(s+t) = b(s)b(t)$ and $b(0) = 1$, hence b has the form $b(t) = e^{tc}$ for a unique complex scalar c.

The positive definiteness of the function $U, V \mapsto k_t(U, V)$ follows from the observation that

$$\sum_{i,j=1}^n \lambda_i \bar{\lambda}_j e^{tc(U_i, U_j)} = A_t^* A_t \geq 0,$$

where $A_t = \lambda_1 U_1(t) + \lambda_2 U_2(t) + \ldots + \lambda_n U_n(t)$. $\qquad\square$

More generally, let X be a nonvoid set and let $c : X \times X \to \mathbb{C}$ be a complex-valued function of two varibles. c is called *conditionally positive definite* if it is self-adjoint in the sense that $\overline{c(x,y)} = c(y,x)$, and obeys

$$\sum_{i,j=1}^n \lambda_i \bar{\lambda}_j c(x_i, x_j) \geq 0$$

for all $x_1, \ldots, x_n \in X$, all $\lambda_1, \ldots, \lambda_n \in \mathbb{C}$ satisfying $\lambda_1 + \ldots + \lambda_n = 0$, and all $n \geq 1$. It is not hard to see that c is conditionally positive definite if, and only if, for every $t \geq 0$ the function $k_t(x,y) = e^{tc(x,y)}$ is positive definite [6]. Such a pair (X, c) will be called an (abstract) covariance function. Covariance functions are the "logarithms" of the positive definite functions.

As it happens, there is only one class of examples that occurs in the theory of E_0-semigroups.

Example 4.3: Let H be a (separable) Hilbert space and let X be the cartesian product $X = \mathbb{C} \times H$. Define $c : X \times X \to \mathbb{C}$ by

$$c((a, \xi), (b, \eta)) = a + \overline{b} + \langle \xi, \eta \rangle.$$

c is clearly self-adjoint, and if $(a_1, \xi_1), \ldots, (a_n, \xi_n) \in X$ and $\lambda_1, \ldots, \lambda_n$ are complex numbers summing to 0, then we have

$$\sum_{i,j=1}^{n} \lambda_i \overline{\lambda}_j (a_i + \overline{a}_j + \langle \xi_i, \xi_j \rangle) = \| \sum_{i=1}^{n} \lambda_i \xi_i \|^2 \geq 0.$$

Hence (X, c) is a covariance function. It is a fact that the covariance function $(\mathcal{U}_\alpha, c)$ associated with an E_0-semigroup α can be put into this form (c.f. [2], Theorem 4.7 and 6.1).

In any case, it is clear from Proposition 4.2 that every E_0-semigroup α, for which $\mathcal{U}_\alpha \neq \emptyset$, gives rise to a canonical covariance function $(\mathcal{U}_\alpha, c)$. We now indicate how, starting with an arbitrary covariance function (X, c), one can construct a Hilbert space $H(X, c)$. Let $\mathbb{C}_0 X$ denote the complex vector space of all functions $f : X \to \mathbb{C}$ satisfying

(i) $f(x) = 0$ for all but finitely many x,

(ii) $\sum_x f(x) = 0$.

We can define a sesquilinear form $\langle \cdot, \cdot \rangle$ on $\mathbb{C}_0 X$ by

$$\langle f, g \rangle = \sum_{x,y} f(x) \overline{g(y)} c(x, y),$$

and $\langle \cdot, \cdot \rangle$ is positive semidefinite because c is conditionally positive definite. Thus

$$\mathcal{N} = \{ f \in \mathbb{C}_0 X : \langle f, f \rangle = 0 \}$$

is a linear subspace of $\mathbb{C}_0 X$ and $\langle \cdot, \cdot \rangle$ induces an inner product on the quotient $\mathbb{C}_0 X / \mathcal{N}$. We define

(4.5) $H(X, c) = \text{ completion of } \mathbb{C}_0 X / \mathcal{N}.$

Thus, $\dim H(X, c)$ defines a numerical invariant associated with any covariance function (X, c). It is a simple matter to calculate the dimension of the covariance functions of the preceding example.

14

Proposition 4.6. *Let H be an n-dimensional Hilbert space and let (X, c) be the co-variance function of Example 4.3. Then* $\dim H(X, c) = n$.

Proof: It is enough to exhibit a unitary operator $U : H(X, c) \to H$, where $X = \mathbb{C}_0 \times H$ and c is defined as in Example 4.3. Define a linear transformation $U_0 : \mathbb{C}_0 X \to H$ by

$$U_0(f) = \sum_{(a,\xi)} f(a, \xi) \cdot \xi \quad .$$

U_0 is surjective because every vector $\xi \in H$ can be written $\xi = U_0(f)$ for

$$f(x) = \begin{cases} +1, & \text{if } x = (0, \xi), \\ -1, & \text{if } x = (0, 0), \\ 0, & \text{otherwise.} \end{cases}$$

Moreover, one checks easily that $\langle U_0(f), U_0(g) \rangle_H = \langle f, g \rangle_{\mathbb{C}_0 X}$. Hence U_0 induces an isometry of the quotient $\mathbb{C}_0 X / \mathcal{N}$ onto H, and the closure of the latter is the required unitary operator. $\qquad\qquad\square$

Now let α be an E_0-semigroup. We define a cardinal number $d_*(\alpha)$ as follows. If $\mathcal{U}_\alpha = \emptyset$ we set $d_*(\alpha) = c$, the cardinality of the continuum. If $\mathcal{U}_\alpha \neq \emptyset$, $d_*(\alpha)$ is defined as the dimension of the Hilbert space $H(\mathcal{U}_\alpha, c)$. It is a fact (though not an obvious one) that in the latter case $H(\mathcal{U}_\alpha, c)$ is always separable ([2], Proposition 5.2). Thus, the possible values of $d_*(\alpha)$ are $\{0, 1, 2, \ldots, \infty, c\}$, where ∞ denotes $\aleph_0$, and we have $d_*(\alpha) = c$ iff $\mathcal{U}_\alpha = \emptyset$. Later on, we will establish the following properties of this numerical index d_*:

(i) $d_*(\alpha) = 0$ iff α is degenerate.

(ii) d_* is an invariant for outer conjugacy.

(iii) If α is the CCR flow of rank n then $d_*(\alpha) = n$.

(iv) $d_*(\alpha \otimes \beta) = d_*(\alpha) + d_*(\beta)$.

§5. Product Systems and E_0-semigroups

The Fredholm index of a semigroup U of isometries is given by

$$\text{index } U = \dim \mathcal{E}_+$$

where $\mathcal{E}_+$ is the deficiency space associated with the generator as in §1. In this section we will introduce a new structure E_α, associated with an arbitrary E_0-semigroup α, which occupies a position analogous to that of $\mathcal{E}_+$. In particular, we will be able to define the *dimension* of a structure of this kind and we will find that

$$d_*(\alpha) = \dim E_\alpha$$

(see 6.2).

Let $\alpha = \{\alpha_t : t \geq 0\}$ be an E_0-semigroup acting on $\mathcal{B}(H)$. For every $t > 0$, let $E_\alpha(t)$ be the following linear space of operators

$$(5.1) \qquad E_\alpha(t) = \{T \in \mathcal{B}(H) : \alpha_t(A)T = TA, \ A \in \mathcal{B}(H)\}.$$

Let $p : E_\alpha \to (0,\infty)$ be the total family of vector spaces, i.e.,

$$E_\alpha = \{(t,T) \in (0,\infty) \times \mathcal{B}(H) : T \in E_\alpha(t)\}$$
$$p(t,T) = t, \quad t > 0.$$

This structure has the following properties.

(5.2(i))
$$E_\alpha \text{ is a standard Borel space, and}$$
$$p : E_\alpha \to (0,\infty) \text{ is a surjective Borel map.}$$

That is to say, if we endow $\mathcal{B}(H)$ with the σ-algebra generated by the weak operator topology, then E_α is a Borel subset of the standard Borel space $(0,\infty) \times \mathcal{B}(H)$ and p has the asserted properties.

We claim now that

$$E_\alpha(t) = p^{-1}(t) \text{ is a Hilbert space}$$
$$\text{relative to the inner product } \langle \cdot, \cdot \rangle$$
(5.2(ii))
$$\text{defined on it by}$$
$$\langle S, T \rangle 1 = T^*S, \quad S, T \in E_\alpha(t).$$

Indeed, by the commutation relation which defines $E_\alpha(t)$ it follows that for any pair of its elements S, T we have $T^*S \in \mathcal{B}(H)' = \mathbb{C} \cdot 1$. Hence T^*S must be a scalar operator and this scalar is positive whenever $S = T \neq 0$. A few moments thought shows that the Hilbert space norm on $E_\alpha(t)$ agrees with the operator norm, and hence $E_\alpha(t)$ is a complete inner product space.

The domain of the inner product function is the Borel set $D = \{(x,y) \in E_\alpha \times E_\alpha : p(x) = p(y)\}$, and one verifies that

(5.2(iii))
$$(x,y) \in D \mapsto \langle x, y \rangle \in \mathbb{C} \text{ is a}$$
$$\text{measurable function.}$$

16

Finally, the following condition is nontrivial, and replaces the notion of local triviality for hermitian vector bundles.

(5.2(iv))

> For every $t_0 > 0$, there is a
>
> Borel isomorphism of families θ
>
> which commutes the diagram

$$
\begin{array}{ccc}
E_a & \xrightarrow{\;\theta\;} & (0,\infty) \times E_\alpha(t_0) \\
{\scriptstyle p}\searrow & & \swarrow {\scriptstyle p} \\
& (0,\infty) &
\end{array}
$$

This means that θ is a Borel isomorphism which restricts to a unitary operator on each fiber. (5.2(iv)) is equivalent to the assertion that there is a sequence of measurable sections $t \mapsto e_n(t) \in E_\alpha(t)$, $n = 1, 2, \ldots$, such that $\{e_1(t), e_2(t), \ldots\}$ is an orthonormal basis for $E_\alpha(t)$ for every $t > 0$. The proof can be found in ([2], Lemma 2.3 of Proposition 2.2).

More generally, a structure $p : E \to (0, \infty)$ satisfying the four conditions of (5.2) will be called a measurable family of Hilbert spaces. We will sometimes write $\{E(t) : t > 0\}$, or simply E, in place of $p : E \to (0, \infty)$.

Using operator multiplication, we can introduce a binary multiplication $x, y \mapsto xy$ in E_α as follows

$$(s, S)(t, T) = (s + t, ST).$$

We have

(5.3(i))

> $(x, y) \in E_\alpha \times E_\alpha \mapsto xy \in E_\alpha$ is a jointly
>
> measurable associative operation which is
>
> bilinear on each fiber $E_\alpha(s) \times E_\alpha(t)$, $\quad s, t > 0$.

Moreover, this operation *acts like tensoring* in the following sense.

(5.3(ii))

> For each $s, t > 0$, there is a unique
>
> bounded operator from the Hilbert space
>
> tensor product $E_\alpha(s) \otimes E_\alpha(t)$ into
>
> $E_\alpha(s + t)$ which carries $x \otimes y$ to xy;
>
> moreover, this operator is unitary.

(5.3(ii)) simply means that (a) $E_\alpha(s + t)$ is spanned by $E_\alpha(s)E_\alpha(t) = \{xy : x \in E_\alpha(s),\ y \in E_\alpha(t)\}$, and (b) for all $x, x' \in E_\alpha(s)$ and all $y, y' \in E_\alpha(t)$ we have

$$\langle xy, x'y' \rangle = \langle x, x' \rangle \langle y, y' \rangle.$$

The properties (a) and (b) are easily proved directly from the definition of E_α (c.f. [2], §2). Any structure having these properties will be called a *product system*, i.e.,

17

Definition 5.4. *A product system is a measurable family of Hilbert spaces* $p : E \to (0, \infty)$, *on which there is defined a measurable associative binary operation which acts like tensoring.*

One should keep several things in mind about product systems. First, the family of Hilbert spaces exists over the open interval $(0, \infty)$, not $[0, \infty)$. The reason for this can be seen in the examples E_α arising from E_0-semigroups α. One can of course define a linear space $E_\alpha(0)$ exactly as we have defined $E_\alpha(t)$ for $t > 0$, but

$$E_\alpha(0) = \{T \in \mathcal{B}(H) : TA = AT, \ A \in \mathcal{B}(H)\}$$

is plainly one-dimensional while, in the generic case, the spaces $E_\alpha(t)$ are infinite-dimensional for all positive t. Thus $t = 0$ is a *singularity*, and it is convenient to omit this value from consideration.

Second, a product system $p : E \to (0, \infty)$ has no topology; it carries a Borel structure plus an algebraic operation. Nevertheless, this combination (Borel structure plus algebraic structure) is known to lead to topological structure in many situations. To illustrate this, recall that a measurable one-parameter unitary group $U = \{U_t : t \in \mathbb{R}\}$ acting on a (separable) Hilbert space is necessarily strongly continuous. Hence there is a topological space associated with all such measurable unitary groups, namely the real line $\mathbb{R}$. The connection is made explicit by the following reformulation of Stone's theorem:

For every measurable one-parameter unitary group $U = \{U_t : t \in \mathbb{R}\}$ *there is a unique representation* π *of the C^*-algebra $C_0(\mathbb{R})$ such that*

$$\pi(\hat{f}) = \int_{-\infty}^{\infty} f(t)U_t dt$$

for all $f \in L^1(\mathbb{R})$. Here, $\hat{f} \in C_0(\mathbb{R})$ denotes the Fourier transform of $f \in L^1(\mathbb{R})$.

Similarly, with every product system E, there is an associated noncommutative topological space $C^*(E)$. More precisely, $C^*(E)$ is a separable C^*-algebra with the property that the representations of E (c.f. (5.8) below) correspond bijectively with the $*$-representations of $C^*(E)$. These C^*-algebras are "continuous" analogues of the Cuntz algebra $\mathcal{O}_\infty$ [5], and their study will be taken up elsewhere.

Finally, it will be convenient to keep in mind the following heuristic picture of a product system E. One thinks of $E(t)$ as a "continuous tensor product" of copies of a single Hilbert space H, taken over the interval $[0, t]$:

(5.5) $$E(t) = \bigotimes_{0 \leq \lambda \leq t} H_\lambda, \qquad H_\lambda = H.$$

However, the germ H often fails to exist. For example, this is precisely the case for product systems E_α arising from E_0-semigroups α with the property that $\mathcal{U}_\alpha = \emptyset$.

We now describe a class of examples for which one *can* make sense out of Formula (5.5). Let K be a Hilbert space of dimension n, $1 \leq n \leq \infty$. Let

$$(5.6) \qquad e^{L^2((0,\infty);K)}$$

denote the symmetric Fock space constructed from the one-particle space $L^2((0,\infty);K)$ of all K-valued L^2-functions on $(0,\infty)$ (c.f. §3). For every $t > 0$, let $L^2((0,t);K)$ denote all functions in $L^2((0,\infty);K)$ which vanish a.e., outside the interval $(0,t)$. Then we may consider

$$(5.7) \qquad E(t) = e^{L^2((0,t);K)}, \quad t > 0$$

to be a subspace of the Hilbert space (5.6). This defines a measurable family of Hilbert spaces $p : E \to (0,\infty)$ via

$$E = \{(t,\xi) : t > 0, \ \xi \in E(t)\}$$

$$p(t,\xi) = t.$$

There is a natural multiplication in E. For example, if $x \in E(s)$ and $y \in E(t)$ are of the form

$$x = \exp f, \quad y = \exp g$$

with $f \in L^2((0,s);K)$, $g \in L^2((0,t);K)$, then xy is defined to be $\exp h$ where

$$h(\lambda) = \begin{cases} f(\lambda), & 0 < \lambda \leq s \\ g(\lambda - s), & s < \lambda \leq s+t \\ 0, & \lambda > s+t. \end{cases}$$

Notice that $h = f + T_s(g)$ is the sum of f and the s-translate of g. One can show that there is unique bounded bilinear map of $E(s) \times E(t)$ into $E(s+t)$ which extends the above definition, and that this multiplication makes $p : E \to (0,\infty)$ into a product system. E is called the *standard example of rank n*. It turns out to be isomorphic to the product system E_α associated to the CCR flow α of rank n.

Now in general, the symmetric Fock space construction $H \to e^H$ has the property that there is a natural identification

$$e^{H \oplus K} \cong e^H \otimes e^K$$

19

(c.f., §3). Thus, direct integrals of Hilbert spaces are turned into "continuous tensor products". In particular, for the standard examples E constructed from the one-particle space $L^2((0, \infty); K)$ as above, one can interpret the Equation (5.7) defining $E(t)$ as a precise formulation of (5.5), i.e.,

$$E(t) = \bigotimes_{0 \leq \lambda \leq t} H, \qquad \text{where } H = e^K.$$

It is possible to make the meaning of this formula more precise, but we will not take the time to do that here.

Returning now to semigroups, let α be an E_0-semigroup and let $p : E_\alpha \to (0, \infty)$ be its associated product system. Notice that the map $\phi : E_\alpha \to \mathcal{B}(H_\alpha)$ defined by

$$\phi(t, T) = T$$

has the following properties.

(i) $\phi : E_\alpha \to \mathcal{B}(H)$ is a measurable function.

(ii) $\phi(xy) = \phi(x)\phi(y)$.

(iii) For every $t > 0$, $\phi|_{E_\alpha(t)}$ is linear and satisfies

$$(5.8) \qquad \phi(y)^* \phi(x) = \langle x, y \rangle 1, \quad x, y \in E_\alpha(t).$$

More generally, a map $\phi : E \to \mathcal{B}(H)$ with the properties (5.8) will be called a *representation* of the product system E. Thus, an E_0-semigroup α acting on $\mathcal{B}(H)$ gives rise to a *pair* (E_α, ϕ), consisting of a product system E_α and a representation $\phi : E_\alpha \to \mathcal{B}(H)$.

Let $\phi : E \to \mathcal{B}(H)$ be a representation of an abstract product system E. For every $t > 0$, let H_t be the following subspace of H,

$$H_t = [\phi(x)\xi : \xi \in H, \ x \in E(t)].$$

The subspaces H_t are *decreasing* in t, and it is a fact that we always have

$$\overline{\bigcup_{t > 0} H_t} = H,$$

([2], corollary of Proposition 2.7). On the other hand, it may well happen that $H_t \neq H$ for all positive t. We distinguish the two extreme cases in which

$$H_t = H, \quad \text{for every } t > 0$$

and in which

$$\bigcap_{t>0} H_t = \{0\}.$$

In the former case ϕ is called *nonsingular* and in the latter ϕ is called *singular*. It is easy to see that every representation ϕ of a product system decomposes uniquely into a central direct sum

$$\phi = \phi_1 \oplus \phi_2$$

where ϕ_1 is nonsingular and ϕ_2 is singular. Finally, a product system E is called *nonsingular* if it admits at least one nonsingular representation.

The product systems that arise from E_0-semigroups are always nonsingular. Indeed, if α is an E_0-semigroup acting on $\mathcal{B}(H)$ and $\phi : E \to \mathcal{B}(H)$ is its natural representation

$$\phi(t, T) = T,$$

then it is quite easy to show that $\alpha_t(1) = 1$ is the projection onto the subspace

$$H_t = [\phi(E_\alpha(t))H].$$

Hence ϕ is a nonsingular representation and so E_α *is a nonsingular product system*. It is conceivable (perhaps even likely) that every product system is nonsingular, but at present we do not have a proof.

In any case, the process of starting with an E_0-semigroup α and constructing out of it a product system E_α and a nonsingular representation $\phi : E_\alpha \to \mathcal{B}(H_\alpha)$ can be turned around as follows. If E is any product system and $\phi : E \to \mathcal{B}(H)$ is any nonsingular representation of E, then there is a unique E_0-semigroup α acting on $\mathcal{B}(H)$ which satisfies the following condition for every positive t:

$$\alpha_t(A)\phi(x) = \phi(x)A, \quad A \in \mathcal{B}(H), \ x \in E(t).$$

See ([2], Proposition 2.7). Moreover, the product system E_α associated to this E_0-semigroup α is isomorphic to E.

We have already alluded to the fact that one can associate a C^*-algebra $C^*(E)$ to every product system E in such a way that representations of E correspond to $*$-representations of $C^*(E)$. Thus the problem of classifying E_0-semigroups α is analogous to the problem of classifying, say, normal operators T. To analyze T one first computes the spectrum of T and then analyzes various properties of the representation

$$f \in C(\sigma(T)) \mapsto f(T).$$

Similarly, if one is presented with an E_0-semigroup α for study, ones first step should be to calculate the "topological" data $C^*(E_\alpha)$, and then attempt to classify the representations of $C^*(E_\alpha)$ under various equivalence relations. In the following section we describe the progress we have made on the second problem. A detailed study of the first problem will be taken up elsewhere.

§6. Outer Conjugacy and the Semigroup Σ

Two product systems E and F are said to be *isomorphic* if there is a Borel isomorphism $\theta : E \to F$ which restricts to a unitary operator on each fiber and satisfies $\theta(xy) = \theta(x)\theta(y)$ for all $x, y \in E$. An *anti-isomorphism* is a similar map which reverses multiplication.

Let Σ be the set of all isomorphism classes of *nonsingular* product systems. We can make Σ into a commutative semigroup as follows. There is a natural way to define the tensor product of two product systems E and F. The fiber space of $E \otimes F$ over $t > 0$ is the tensor product of Hilbert spaces $E(t) \otimes F(t)$, and multiplication in $E \otimes F$ is uniquely defined by specifying that elementary tensors should multiply in the obvious way. If E and F are nonsingular then so is $E \otimes F$. Thus we may define "addition" in Σ by

$$[E] + [F] = [E \otimes F],$$

and this makes Σ into an abelian semigroup.

This semigroup has a zero element which is defined as follows. Let $p : Z \to (0, \infty)$ be the *trivial* product system, defined as follows:

$$Z = (0, \infty) \times \mathbb{C}$$
$$p(t, z) = t.$$

The inner product in each fiber $Z(t)$ is the usual one in $\mathbb{C}$, i.e., $\langle z, w \rangle = z\overline{w}$, and multiplication is defined by

$$(s, z)(t, w) = (s + t, zw).$$

Z is nonsingular because the map $\phi : Z \to \mathbb{C}$ given by

$$\phi(t, z) = z$$

is a one-dimensional nonsingular representation of Z. One may verify easily that

$$E \cong Z \otimes E$$

for every product system E, and hence the element of Σ defined by

$$0 = [Z]$$

functions as an additive zero.

We now define a numerical invariant for product systems which generalizes what was done in §4 for E_0-semigroups. Let E be a product system. A *unit* of E is a measurable cross section $t \in (0, \infty) \mapsto u_t \in E(t)$ which is not identically zero and is multiplicative

$$u_{s+t} = u_s u_t, \quad s, t > 0.$$

$\mathcal{U}_E$ will denote the set of units of E. Notice that every unit u of E gives rise to a homomorphism ω_u of E into the trivial product system Z as follows,

$$\omega_u(x) = \langle x, u_t \rangle, \quad x \in E(t), \; t > 0.$$

More precisely, ω_u is a measurable function from E to Z which is linear on fibers, multiplicative, and is not identically zero. Conversely, it is not hard to show that every nonzero element ω in $\hom(E, Z)$ arises in this way from a unique unit in $\mathcal{U}_E$. Thus, one may think of the set $\mathcal{U}_E$ of units as a counterpart of the Gelfand spectrum of a commutative Banach algebra. Again, we remind the reader that $\mathcal{U}_E$ can be empty.

Proposition 6.1. *For any two units u, v of a product system E there is a unique complex number $c(u, v)$ such that*

$$\langle u_t, v_t \rangle = e^{tc(u,v)}, \quad t > 0.$$

Of course, Proposition 6.1 is similar to Proposition 4.2, though its proof is somewhat more delicate ([2], Theorem 4.1). The function $c : \mathcal{U}_E \times \mathcal{U}_E \to \mathbb{C}$ is conditionally positive definite and is called the *covariance function* of E.

Apparently, we can proceed exactly as in §4 to construct a Hilbert space $H(\mathcal{U}_E, c)$ whenever $\mathcal{U}_E \neq \emptyset$, and we may define the *dimension* of an arbitrary product system E in the obvious way

$$\dim E = \begin{cases} \dim H(\mathcal{U}_E, c), & \text{if } \mathcal{U}_E \neq \emptyset, \\ c, & \text{if } \mathcal{U}_E = \emptyset. \end{cases}$$

It is clear that $\dim E$ depends only on the isomorphism class of E and so defines a map

$$\dim_* : \Sigma \to \{0, 1, 2, \ldots, \infty, c\}.$$

Now suppose that $E = E_\alpha$ is the product system associated with an E_0-semigroup α. There is an apparent discrepancy between the definition of unit given in §4 and the definition given above; the former refers to a strongly continuous semigroup defined on $[0, \infty)$ while the latter refers to a (weakly) measurable semigroup defined on $(0, \infty)$. It is a nontrivial fact that these two definitions agree [2], and in fact we have the formula promised at the beginning of §5:

$$(6.2) \qquad\qquad d_*(\alpha) = \dim E_\alpha.$$

The purpose of this section is to show that the abstract Powers–Robinson index actually takes its values in the semigroup Σ, and to describe the role played by product systems in the problem of classifying E_0-semigroups to outer conjugacy.

We first recall two definitions from [9]. Two E_0-semigroups α, β are said to be *paired* (written $(\alpha, \beta) \in \mathcal{P}$) if there is a one-parameter group $\gamma = \{\gamma_t : t \in \mathbb{R}\}$ of $*$-automorphisms of $\mathcal{B}(H_\alpha \otimes H_\beta)$ such that

$$\gamma_t(A \otimes 1) = \alpha_t(A) \otimes 1,$$
$$\gamma_{-t}(1 \otimes B) = 1 \otimes \beta_t(B), \quad t \geq 0$$

for all $A \in \mathcal{B}(H_\alpha)$, $B \in \mathcal{B}(H_\beta)$. It is still unknown if pairing is an equivalence relation. The relation is clearly symmetric, but it is not known to be transitive or even if every α can be paired with itself. α and β are said to have the *same index* if there is a third E_0-semigroup σ such that

$$(\alpha, \sigma) \in \mathcal{P} \quad \text{and} \quad (\sigma, \beta) \in \mathcal{P}.$$

Powers and Robinson showed that *same index* is an equivalence relation, that CAR flows of different ranks are inequivalent, and that the example of [8] is not equivalent to any CAR flow.

Our first result characterizes pairing in terms of product systems ([2], Theorem 3.4).

Theorem 6.3. *Two E_0-semigroups α, β are paired iff their associated product systems E_α, E_β are anti-isomorphic.*

In particular, $(\alpha, \alpha) \in \mathcal{P}$ iff E_α is anti-isomorphic to itself. While this explains, perhaps, why it is that one runs into difficulty in trying to prove that pairing is an equivalence relation, we do not know an example of a product system which is not anti-isomorphic to itself.

24

Theorem 6.4. *Two E_0-semigroups α, β are outer conjugate iff their associated product systems E_α, E_β are isomorphic.*

This is proved in ([2], corollary of Theorem 3.18). In particular, it shows that if one wants to classify E_0-semigroups to outer conjugacy, one should attempt to classify product systems up to isomorphism. While this program has enjoyed some success (c.f., Theorem 6.5 below), the exceptional product systems of dimension c remain quite mysterious. Ultimately, one would like to know the structure of Σ. At this point, we do not even know its *cardinality*. Note, for example, that the results of §7 show that there is a subjective homomorphism of Σ onto the abelian semigroup $\{0, 1, 2, \ldots, \infty, c\}$.

Finally, because of the formula

$$d_*(\alpha) = \dim E_\alpha,$$

Theorem 6.4 implies that *the index of an E_0-semigroup is stable under outer conjugacy*, and in particular under bounded perturbations of its generator.

Here is a simple illustration of how one applies Theorem 6.4.

Corollary. *Let α, β be E_0-semigroups and suppose that each β_t is an automorphism of $\mathcal{B}(H_\beta)$. Then $\alpha \otimes \beta$ is outer conjugate to α.*

Proof: It suffices to show that the product systems $E_{\alpha \otimes \beta}$ and E_α are isomorphic. Now $E_{\alpha \otimes \beta} \cong E_\alpha \otimes E_\beta$ (c.f. [2], Proposition 3.15) and, since β is degenerate, its product system has one-dimensional fiber spaces $E_\beta(t)$, $t > 0$. By the corollary of Lemma 7.3 below, we have $E_\beta \cong Z$ and hence $E_{\alpha \otimes \beta} \cong E_\alpha \otimes Z \cong E_\alpha$. $\square$

Together, Theorems 6.3 and 6.4 combine to establish the equivalence of the following three assertions about E_0-semigroups α, β:

(i) α and β have the same Powers–Robinson index.

(ii) α and β are outer conjugate.

(iii) E_α and E_β are isomorphic.

We conclude this section by describing the classification results we have been able to obtain so far. These results follow from a detailed analysis of the product systems associated with E_0-semigroups (see [2], §§6–7). Let α be an E_0-semigroup acting on $\mathcal{B}(H)$. For every $t > 0$, we can form the closed linear subspace D_t of H generated by all vectors of the form

$$U_1(s_1)U_2(s_2)\ldots U_n(s_n)\xi,$$

where $\xi \in H$, $U_1, \ldots, U_n \in \mathcal{U}_\alpha$, and $s_1, \ldots, s_n$ are positive numbers *summing to t*. α is said to be *strongly spatial* if $D_t = H$ for some (and therefore every) positive t. This means that $\mathcal{U}_\alpha$ has "sufficiently many" elements. In [2], this property was called *spatial*; we have changed the terminology because Powers has already used the term *spatial* to describe the property $\mathcal{U}_\alpha \neq \emptyset$.

Theorem 6.5. *Let α be a strongly spatial E_0-semigroup and let $n = d_*(\alpha)$. Then α is outer conjugate to the CCR flow of rank n.*

In particular, we may conclude that strongly spatial E_0-semigroups α are completely classified to outer conjugacy by their index $d_*(\alpha)$. The proof of Theorem 6.5 proceeds by showing that, under the stated hypotheses, the product system E_α is isomorphic to the standard example of

§7. The Addition Formula

In this section we discuss the logarithmic addition formula

$$(7.1) \qquad\qquad d_*(\alpha \otimes \beta) = d_*(\alpha) + d_*(\beta)$$

for the numerical index of E_0-semigroups α, β. In view of the relation $d_*(\alpha) = \dim E_\alpha$, it suffices to prove a slightly more general formula relating the dimension of product systems $E, F, E \otimes F$:

$$(7.2) \qquad\qquad \dim E \otimes F = \dim E + \dim F.$$

Our proof of (7.2) makes essential use of the theory of product systems, together with the following

Lemma 7.3. *Every multiplier of $(0, \infty)$ is trivial.*

The meaning of Lemma 7.3 is as follows. A Borel function

$$m : (0, \infty) \times (0, \infty) \to \{|z| = 1\}$$

is called a *multiplier* if it satisfies the functional equation

$$m(x, y + z)m(y, z) = m(x + y, z)m(x, y)$$

for all $x, y, z > 0$. A multiplier m is called *trivial* if there is a Borel function $f : (0, \infty) \to \{|z| = 1\}$ such that

$$m(x, y) = \frac{f(x + y)}{f(x)f(y)}, \quad x, y > 0.$$

It is well known that every multiplier on $\mathbb{R} \times \mathbb{R}$ is trivial. But we have been unable to deduce Lemma 7.3 from that result in a straightforward way. Our original proof of Lemma 7.3 used an extension procedure for semigroups of endomorphisms of the C^*-algebra of compact operators [3]. Subsequently, Paul Chernoff and Palle Jorgensen have each given a proof along different lines.

We digress to deduce a consequence of Lemma 7.3 which gives a strong uniqueness property of the trivial product system Z. If $\{E(t) : t > 0\}$ is any family of Hilbert spaces with the property that $E(s + t)$ is isomorphic to $E(s) \otimes E(t)$ for all $s, t > 0$, then the dimension $d(t)$ of the fiber spaces $E(t)$ must satisfy the functional equation $d(s+t) = d(s)d(t)$, and hence d is identically 1 or ∞. In the former case, it is conceivable that there are nontrivial product systems $p : E \to (0, \infty)$ satisfying $\dim E(t) = 1$ for all t. The following corollary rules this out.

Corollary. *If E is a product system satisfying $\dim E(t) = 1$ for all $t > 0$, then $E \cong Z$.*

Proof: We may choose a measurable section of unit vectors $e_t \in E(t)$, $t > 0$. Fix $s, t > 0$. Because $E(s + t)$ is one-dimensional, there is a unique scalar $m(s, t)$ satisfying

$$e_{s+t} = m(s, t)e_s e_t.$$

Clearly $|m(s, t)| = 1$, and a simple argument using the associative law for multiplication in E shows that m is a multiplier. Lemma 7.3 implies that m has the form

$$m(s, t) = \frac{f(s + t)}{f(s)f(t)}$$

for a measurable function $f : (0, \infty) \to \{|z| = 1\}$, and hence we may define a unit u of E by

$$u_t = f(t)^{-1}e_t.$$

As we have seen before, the map $\omega : E \to Z$ defined by

$$\omega(x) = \langle x, u_t \rangle, \quad x \in E(t)$$

is a homomorphism of product systems which in this case is clearly unitary on fiber spaces. Hence ω is an isomorphism. $\qquad \square$

We turn now to the formula (7.2). This means basically that if $\mathcal{U}_{E \otimes F} \neq \emptyset$, then both $\mathcal{U}_E$ and $\mathcal{U}_F$ are nonvoid and moreover

$$(7.4) \qquad H(\mathcal{U}_{E \otimes F}, c_{E \otimes F}) \cong H(\mathcal{U}_E, c_E) \oplus H(\mathcal{U}_F, c_F).$$

Now if we are given units $u \in \mathcal{U}_E$, $v \in \mathcal{U}_F$, then we can form a unit $u \otimes v \in \mathcal{U}_{E \otimes F}$ in the obvious way

$$(u \otimes v)_t = u_t \otimes v_t, \quad t > 0.$$

Notice that

$$(7.5) \qquad c_{E \otimes F}(u \otimes v, u' \otimes v') = c_E(u, u') + c_F(v, v').$$

Indeed, if we multiply the left side of (7.5) by $t > 0$ and exponentiate we obtain $\langle u_t \otimes v_t, u'_t \otimes v'_t \rangle$. Doing the same thing to the right side leads to

$$\langle u_t, u'_t \rangle \langle v_t, v'_t \rangle = \langle u_t \otimes v_t, u'_t \otimes v'_t \rangle,$$

and (7.5) follows after an obvious differentiation.

Now because of (7.5), the mapping $\theta : \mathcal{U}_E \times \mathcal{U}_F \to \mathcal{U}_{E \otimes F}$ defined by $\theta(u, v) = u \otimes v$ induces a linear isometry

$$\hat{\theta} : H(\mathcal{U}_E, c_E) \oplus H(\mathcal{U}_F, c_F) \hookrightarrow H(\mathcal{U}_{E \otimes F}, c_{E \otimes F}).$$

We need to show that $\hat{\theta}$ is surjective. This will follow if we show that θ is surjective, i.e.,

Theorem 7.6. *Let E, F be product systems. Then every unit w of $E \otimes F$ can be decomposed $w = u \otimes v$ where $u \in \mathcal{U}_E$, $v \in \mathcal{U}_F$.*

In particular, Theorem 7.6 implies that if $\mathcal{U}_{E \otimes F} \neq \emptyset$ then both $\mathcal{U}_E$ and $\mathcal{U}_F$ are nonvoid. We will describe the idea behind the proof of Theorem 7.6. By a *morphism* of product systems we mean a Borel map $\theta : E \to F$ such that

(i) $\theta(xy) = \theta(x)\theta(y)$, $x, y \in E$, and

(ii) For each $t > 0$, $\theta_t = \theta|_{E(t)}$ is a bounded linear operator from $E(t)$ to $F(t)$.

A morphism $\theta : E \to F$ is called *compact* if θ_t is a compact operator for every $t > 0$. Here are two examples.

Example 7.7: Let $u \in \mathcal{U}_E$, $v \in \mathcal{U}_F$ and define $\theta : E \to F$ by

$$\theta(x) = \langle x, u_t \rangle v_t, \quad x \in E(t), \ t > 0.$$

θ is a morphism because u and v are units, and θ_t is rank-one for every positive t.

Example 7.8: Let E, F be product systems and let w be a unit of $E \otimes F$. For fixed $t > 0$, consider the bounded bilinear form

$$(x, y) \in E(t) \times F(t) \mapsto \langle x \otimes y, w_t \rangle.$$

Let $\overline{F}$ be the product system *conjugate* to F. This is to say that the product system $\overline{F}$ is identical to F except that scalar multiplication in the fiber spaces is conjugated. Thus, if $x \in \overline{F}(t)$ and $\lambda \in \mathbb{C}$, then $\lambda \cdot x$ means $\overline{\lambda} x$. The natural image of an element x of F in the conjugate system $\overline{F}$ is denoted $\overline{x}$. Then $x \mapsto \overline{x}$ is multiplicative and anti-unitary on fiber spaces. So by the Riesz lemma we can assert that for every $t > 0$ there is a unique bounded linear operator $\theta_t : E(t) \to \overline{F}(t)$ such that

(7.9) $$\langle \theta_t(x), \overline{y} \rangle = \langle x \otimes y, w_t \rangle$$

for all $x \in E(t)$, $y \in F(t)$. It is obvious that $\theta : E \to \overline{F}$ is a morphism of product systems. Moreover, a few moments' reflection upon (7.9) shows that each θ_t is a *Hilbert Schmidt* operator.

Now the essential content of Theorem 7.6 is simply that *every θ of Example 7.8 is actually of the simpler form of Example 7.7*. Therefore, 7.6 will follow from

Theorem 7.10. *For every nonzero compact morphism $\theta : E \to F$, there exist units $u \in \mathcal{U}_E$, $v \in \mathcal{U}_F$ such that θ has the form*

$$\theta(x) = \langle x, u_t \rangle v_t, \quad x \in E(t), \ t > 0.$$

The proof of Theorem 7.10 can be found in ([3], Theorem 3.4).

It follows from (7.2) that the dimension function induces a surjective homomorphism of abelian semigroups

$$\dim_* : \Sigma \to \{0, 1, 2, \ldots, \infty, c\}.$$

The structure of the "kernel" of this map is unknown.

§8. Calculation of $d_*(\alpha)$

Until now, we have said nothing about how one goes about calculating the numerical index $d_*(\alpha)$ of an E_0-semigroup α. Our principal result along these lines is summarized as follows.

Theorem 8.1. *Let $U = \{U_t : t \geq 0\}$ be a semigroup of isometries and let α^U be the E_0-semigroup constructed from U as in §3. Then*

$$d_*(\alpha^U) = \ index\ U.$$

Because of the properties of the functor $U \to \alpha^U$, the proof of 8.1 reduces to the case where U is the shift of multiplicity n,

$$(8.2) \qquad U_t f(x) = \begin{cases} f(x-t), & x > t \\ 0, & 0 < x \leq t \end{cases},$$

$f \in L^2((0, \infty); K)$, K being a Hilbert space of dimension n. In this case, one can show directly that the product system $E(\alpha^U)$ is isomorphic to the *standard* product system $E_K = \{E(t) : t > 0\}$ associated to K as in equation (5.7). Thus the proof of 8.2 is reduced to the problem of calculating the covariance function of the standard examples E_K of rank $n = \dim K$. Here is the result we need.

Theorem 8.3. *Let K be a Hilbert space and let E_K be the product system defined by (5.7). For every pair $(a, \xi) \in \mathbb{C} \times K$, let $u_{(a,\xi)}$ be the section of E_K defined by*

$$u_{(a,\xi)}(t) = e^{at} \exp(\chi_{(0,t)} \otimes \xi), \qquad t > 0.$$

Then $(a, \xi) \mapsto u_{(a,\xi)}$ defines a bijection of $\mathbb{C} \times K$ onto $\mathcal{U}(E_K)$. Moreover, the covariance function c of E_K satisfies

$$c(u_{(a,\xi)}, u_{(b,\eta)}) = a + \bar{b} + \langle \xi, \eta \rangle.$$

The proof of Theorem 8.3 can be found in ([2], Theorem 4.7). Thus the covariance function of E_K is of the type described in Example 4.3. By the computation of Proposition 4.6 we conclude that

$$\dim(E_K) = \dim K,$$

and since $\dim K$ is the Fredholm index of the semigroup U of 8.2, we obtain Theorem 8.1.

References

1. Araki, H. and Woods, E. J., Complete Boolean algebras of type I factors, Publ. Res. Inst. Math. Sciences, Kyoto University Series A, Vol. II (1966), 157–242.
2. Arveson, W., Continuous analogues of Fock space, to appear.

3. Arveson, W., An addition formula for the index of semigroups of endormophisms of $\mathcal{B}(H)$, Pac. J. Math., to appear.

4. Connes, A., Une classification des facteurs de type III, Ann. Sci. Ecole Norm. Sup., Paris (4) **6** (1973), 133–252.

5. Cuntz, J., Simple C^*-algebras generated by isometries, Comm. Math. Phys. **57** (1977), 173–185.

6. Guichardet, A., Symmetric Hilbert spaces and related topics, Springer–Verlag Lecture Notes in Mathematics, No. 261, Berlin (1972).

7. Powers, R. T., An index theory for semigroups of endomorphisms of $\mathcal{B}(H)$ and type II_1 factors, Canad. J. Math., to appear.

8. Powers, R. T., A non-spatial continuous semigroup of $*$-endomorphisms of $\mathcal{B}(H)$, Publ. Res. Inst. Math. Sciences, Kyoto University, to appear.

9. Powers, R. T., and Robinson, D., An index for continuous semigroups of $*$-endomorphisms of $\mathcal{B}(H)$, J. Func. Anal., to appear.

10. Pedersen, G. K., *C^*-algebras and Their Automorphism Groups*, Academic Press, London (1979).

11. Reed, M. and Simon, B., Methods of mathematical physics, Vol. 11, Academic Press, New York (1975).

William Arveson

Department of Mathematics

University of California

Berkeley, California 94720

Research supported in part by NSF Grant DMS-86-00375.

P. G. Ghatage

Integral representations of some Hankel operators on the Hardy and Bergman spaces

§1. Introduction

In his expository paper S. Power [5] gave an integral representation of positive self–adjoint Hankel operators on the Hardy space of the circle and asked if such a representation was possible for all Hankel operators. That it is so, both for Hardy and Bergman spaces of the upper half plane is implicit in the work of R. Rochberg, based on the representation theorems of R. Coifman and R. Rochberg [see 7]. In this note, we start with the disk and carry the construction through, thus writing down matrix representations of certain integral operators with respect to the orthonormal basis consisting of Lagurre polynomials.

§2. Notation

If

$$\varphi(e^{i\theta}) = \sum_{n \geq 0} a_n e^{in\theta}$$

belongs to $H^2(T)$ then we define the Hankel operator $S_\varphi : H^2(T) \to H^2(T)$ by $S_\varphi f = P(\varphi(e^{i\theta})f(e^{-i\theta}))$, where P is the orthogonal projection of $\mathcal{L}^2(T)$ onto $H^2(T)$. It is well–known [8] that S_φ is bounded (compact) exactly when $\varphi \in BMO$ ($\in VMO$) and $\|S_\varphi\| \sim \|\varphi\|_{BMO}$. More generally if

$$\varphi(z) = \sum_{n \geq 0} a_n z^n$$

with

$$\sum |a_n|^2/(n+1) < \infty,$$

then we define the Hankel operator $H_\varphi : \mathcal{L}_a^2(\mathbf{D}) \to \mathcal{L}_a^2(\mathbf{D})$ by $H_\varphi f = \mathbb{P}(f(\overline{z})\varphi(z))$ where $\mathbb{P}$ is the orthogonal projection of $\mathcal{L}^2(\mathbf{D})$ (with the normalized area measure) onto $\mathcal{L}_a^2$, the space of analytic elements in $\mathcal{L}^2(\mathbf{D})$. It is known that H_φ is bounded (compact) if and only if φ belongs to the Bloch space $\mathcal{B}$ (little Bloch space $\mathcal{B}_0$) and $\|\varphi\|_\mathcal{B} \sim \|H_\varphi\|$. For our immediate use we simply note that $BMO \subseteq \mathcal{B}$ and $VMO \subseteq \mathcal{B}_0$; both inclusions being proper. Also $\mathcal{B}$ can be visualized as the dual of $\mathcal{L}_a^1$, the space of analytic functions on $\mathbf{D}$ which are integrable with respect to the area measure.

On the real line,

$$H^2(\mathbf{R}) = \left\{ \frac{1}{1-it} g\left(\frac{1+it}{1-it}\right), \quad g \in H^2(T) \right\}$$

which is known to be the subspace of $\mathcal{L}^2(\mathbf{R})$ consisting of functions whose Fourier transforms vanish on $(-\infty, 0)$ [1]. The Bergman space A^2 of the upper–half–plane consists of analytic functions which are square–integrable with respect to the area measure. They all have boundary value distributions. There is an orthonormal basis consisting of functions whose boundary values are well–defined functions. Thus for f belonging to A^2, we may write $\hat{f}(t)$ for the Fourier transform of the boundary values of f and note that analyticity of f implies $\hat{f}(t) \equiv 0$ on $(-\infty, 0)$. For a detailed description of this and more general Bergman spaces see [7].

If $\psi \in BMO(\mathbf{R})$ we define the Hankel operator $S_\psi : H^2(\mathbf{R}) \to H^2(\mathbf{R})$ as follows: $S_\psi f = P[f(-x)\psi(x)]$ where P is the orthogonal projection of $\mathcal{L}^2(\mathbf{R})$ onto $H^2(\mathbf{R})$. For a definition of $BMO(\mathbf{R})$ see [2]. If b is an analytic function on the upper–half–plane we define Hankel operator H_b on A^2 as follows: $H_b f = \mathbf{P}[b(w)f(-\overline{w})]$ where $f \in A^2$ and $\mathbf{P}$ is the orthogonal projection of $\mathcal{L}^2(u.h.p.)$ onto A^2, u.h.p being the standard abbreviation for the upper–half–plane.

§3. Theorems on $H^2(\mathbf{R})$

We state the next lemma for completeness and refer the reader to [6, Chapter 2].

Lemma 1. *For*

$$\varphi(z) = \sum_{n \geq 0} a_n z^n \in BMOA \quad (\equiv BMO \cap H^2)$$

let

$$\psi(t) = \sum_{n \geq 0} a_n \left(\frac{1+it}{1-it}\right)^{n+1}.$$

Then $S_\varphi \in \mathcal{B}(H^2(T))$ is unitarily equivalent to $S_\psi \in \mathcal{B}(H^2(\mathbf{R}))$.

The proof simply uses the fact that the canonical map of the circle T onto $\mathbf{R}$ transforms the standard orthonormal basis $\{e^{in\theta}, n \geq 0\}$ of $H^2(T)$ onto the orthonormal basis

$$\left\{ \frac{1}{\sqrt{\pi}(1-it)} \left(\frac{1+it}{1-it} \right)^n, \quad n \geq 0 \right\}$$

of $H^2(\mathbb{R})$. With respect to these two bases, the matrix of S_φ as well as S_ψ is $[a_{i+j}]$.

Lemma 2. *If $\psi \in BMOA(\mathbb{R})$ defined as in Lemma 1 and S_ψ is the Hankel operator on $H^2(\mathbb{R})$, then S_ψ is unitarily equivalent to the integral operator S_k on $\mathcal{L}^2(0,\infty)$ defined by $(S_k f)(t) = \int_0^\infty k(s+t)f(s)ds$ where $k(t) = \hat{\psi}(t)$.*

Proof: If $\mathcal{F} : H^2(\mathbb{R}) \to \mathcal{L}^2(0,\infty)$ is the unitary transformation given by the Fourier transform for $t > 0$,

$$\mathcal{F}(f(-x)\psi(x))(t) = \int_{-\infty}^\infty \hat{\psi}(t-s)f(-x)\hat{}(s)ds$$

$$= \int_{-\infty}^0 \hat{\psi}(t-s)\hat{f}(-s)ds$$

as $\hat{f} \equiv 0$ on $(-\infty, 0) = \int_0^\infty \hat{\psi}(t+s)\hat{f}(s)ds$. Now we simply note that

$$\mathcal{F}[P(f(-x)\psi(x))](t) = \chi_{(0,\infty)}\mathcal{F}(f(-x)\psi(x))(t).$$

Thus the proof is a simple consequence of the convolution theorem and the fact that whenever $\psi \in BMO(\mathbb{R})$ we have,

$$\int_{-\infty}^\infty \frac{|\psi(x)|}{1+x^2}dx < \infty$$

[2, p. 224] and hence $\hat{\psi}(t)$ exists (as a distribution). We note that via Fourier transforms, the orthonormal basis

$$\left\{ \frac{1}{\sqrt{\pi}(1-it)} \left(\frac{1+it}{1-it} \right)^n, \quad n \geq 0 \right\}$$

is mapped into $\{e^{-t}\mathcal{L}_n(2t), \ n \geq 0\}$ where $\mathcal{L}_n$ is the nth Lagurre polynomial [10, Chapter 5]. The proof is a simple application of the residue theorem.

Remarks. (1). Restricting the choice of ψ as defined in Lemma 1 is necessary in view of the fact that the constant function 1 clearly induces the zero Hankel operator on $H^2(\mathbb{R})$ and as a distribution, $\hat{\Lambda}_1\varphi = \int_R \hat{\varphi}(x)dx$ for all φ in the Schwartz space.

(2). If S_k is a bounded operator on $\mathcal{L}^2(0,\infty)$ and the kernel k has the 'Hankel property' with respect to the Lagurre polynomials, i.e., if

$$\int_0^\infty \int_0^\infty k(s+t)e^{-(s+t)}\mathcal{L}_n(2s)\mathcal{L}_m(2t)dsdt = c_{n+m}$$

then S_k is unitarily equivalent to $S_\varphi \in \mathcal{B}(H^2(T))$, where

$$\varphi(z) = \sum_{n \geq 0} c_n z^n \in BMOA.$$

§4. Theorems A^2(u.h.p.)

If $F : D \to$ u.h.p. is the canonical map, $F(z) = i\frac{1+z}{1-z} = w$, then we have a unitary transformation $V : \mathcal{L}^2(\text{u.h.p.}) \to \mathcal{L}^2(D)$ given by

$$(Vf)(z) = \frac{c}{(1-z)^2} f\left(i\frac{1+z}{1-z}\right)$$

which clearly takes A^2 onto $\mathcal{L}^2_a(D)$;

$$(V^{-1}g)(w) = \frac{c}{(w+i)^2} g\left(\frac{w-i}{w+i}\right)$$

and we have the standard orthonormal basis of

$$A^2 : \left\{ \frac{c\sqrt{n+1}}{(w+i)^2} \left(\frac{w-i}{w+i}\right)^n, \quad n \geq 0 \right\}.$$

We recall that the Bergman kernel for the u.h.p. is

$$k_\zeta(w) = \frac{c}{(w-\bar\zeta)^2}.$$

Starting with a function f in the linear span of this standard basis, it is easy to see that if $\hat{f}(t)$ is the Fourier transform of the boundary value function then $f(u+iv)\hat{}(t) = e^{-vt}\hat{f}(t)$, [1] and

$$\int_{-\infty}^{\infty} |f(u+iv)|^2 \, du = \int_0^\infty |f(u+iv)\hat{}(t)|^2 \, dt$$

$$= \int_0^\infty e^{-2vt} |\hat{f}(t)|^2 \, dt \Rightarrow \int_0^\infty \int_{-\infty}^\infty |f(u+iv)|^2 \, du \, dv$$

$$= \int_0^\infty e^{-2vt} |\hat{f}(t)|^2 \, dt \, dv = \frac{1}{2} \int_0^\infty t^{-1} |\hat{f}(t)|^2 \, dt.$$

This induces a unitary operator from A^2 onto $\mathcal{L}^2(0,\infty)$ which takes a function $f \mapsto ct^{-\frac{1}{2}}\hat{f}(t)$ where $\hat{f}(t)$ is the Fourier transform of the boundary value distribution (which is known to exist).

36

In keeping with the definition of $BMO(\mathbb{R})$ [2], we define Bloch space of the u.h.p. as the space of analytic functions which are mapped into Bloch space of the disk, when composed with the canonical map F. Explicitly, f belongs to the Bloch space of the u.h.p. if f is analytic and $\sup_{v>0} v|f'(w)| < \infty$ where $w = u + iv$. It follows easily from the corresponding result for functions on the disk that $\sup_{v>0} v^k|f^k(w)| < \infty \forall k \geq 1$. This definition is more restrictive than the one in [7] but has the advantage of not having to give special consideration to functions like z, $z^2 \cdots$. Following Lueking's techniques [4] almost verbatim, any function f in the Bloch space of $\mathbb{D}$ can be written as

$$f = \sum_{n \geq 1} c_n k_{\lambda_n} / \|k_{\lambda_n}\|_2^2$$

where k_λ is the Bergman kernel for $\mathbb{D}$ (with $\sup |c_n| \sim \|f\|_{\mathcal{B}}$ and the series converging to f in the wk. $*$ topology) for a suitably chosen sequence $\{\lambda_n\} \subseteq \mathbb{D}$ (independent of f). This transfers easily to the u.h.p. and any $g \in \mathcal{B}(\text{u.h.p.})$ can be written as

$$g(w) = \sum c_n \frac{v_n^2}{|w_n + i|^4} \frac{1}{(w - \overline{w}_n)^2}$$

where $w_n = u_n + iv_n$ is a suitably chosen sequence in u.h.p.; $\|g\|_{\mathcal{B}} \sim \sup |c_n|$ and series converging to g in the wk. $*$ topology.

Proposition 1. *If*

$$\varphi(z) = \sum_{n \geq 0} a_n z^n \in \mathcal{B},$$

then the Hankel operator $H_\varphi \in \mathcal{B}\left(\mathcal{L}_a^2(\mathbb{D})\right)$ is unitarily equivalent to the Hankel operator $H_b \in \mathcal{B}(A^2)$ where

$$b(w) = \sum_{n \geq 0} a_n \left(\frac{w - i}{w + i}\right)^{n+2},$$

$w \in$ *u.h.p.*

Proof: Note first that if φ belongs to the Bloch space of the disk then b belongs to the Bloch space of the u.h.p. and both the Hankel operators are bounded [7, 8]. If V is the map from $\mathcal{L}^2(\text{u.h.p}) \to \mathcal{L}^2(\mathbb{D})$ (described at the beginning of the section), then for $g \in \mathcal{L}_a^2(\mathbb{D})$ we have,

$$V H_b V^{-1} g = V \mathbb{P}\left[(V^{-1}g)(-\overline{w})b(w)\right] = V \mathbb{P}\left[\frac{2i}{(-\overline{w} + i)^2} g\left(\frac{-\overline{w} - i}{-\overline{w} + i}\right) b(w)\right]$$

$$= P\left[g(\overline{z})\frac{(1 - \overline{z})^2}{(1 - z)^2} b\left(i\frac{1 + z}{1 - z}\right)\right] = H_\varphi g$$

where

$$\varphi = P\left[\frac{(1-\bar{z})^2}{(1-z)^2}b\left(i\frac{1+z}{1-z}\right)\right].$$

Note that

$$P\left[z^\ell\left(\frac{1-\bar{z}}{1-z}\right)^2\right] = z^{\ell-2}$$

for $\ell \geq 2$, and that

$$b\left(i\frac{1+z}{1-z}\right)$$

is analytic on $\mathbf{D}$. This establishes the relation between φ and b.

Proposition 2. *If b is the analytic function described in Proposition 1, then H_b is unitarily equivalent to the integral operator K_b on $\mathcal{L}^2(0,\infty)$ defined by*

$$(K_b f)(t) = \int_0^\infty \frac{\sqrt{st}}{s+t}\hat{b}(s+t)f(s)ds$$

where $\hat{b}(t)$ denotes the Fourier transform of the boundary value distribution of b.

Proof: We recall that

$$A^2 \to \mathcal{L}^2(0,\infty) \quad f \overset{\mathcal{F}_1}{\to} c\hat{f}(t)t^{-\frac{1}{2}}$$

is an isometry. The easiest way to see that it is indeed onto is to note that under this mapping, the standard orthonormal basis of A^2 is transformed into

$$\left\{c\frac{e^{-t}t^{1/2}}{\sqrt{n+1}}\mathcal{L}_n^1(2t), \quad n \geq 0\right\}$$

where $\mathcal{L}_n^1$ is n^{th} Lagurre polynomial of order 1. It is well–known that this is a complete orthonormal set in $\mathcal{L}^2(0,\infty)$. See [10, Chapter 5]. Now if f in A^2 belongs to the linear span of its standard orthonormal basis then

$$(H_b f)(\zeta) = P\left[f(-\bar{w})b(w)\right](\zeta) = c\int_0^\infty\int_{-\infty}^\infty f(-\bar{w})b(w)\frac{1}{(\bar{w}-\zeta)^2}du\,dv$$

where $\zeta = \xi + i\eta; (\eta > 0)$.

38

For $t > 0$,

$$(\mathcal{F}_1 H_b f)(t) = ct^{-1/2} \int_0^\infty \int_{-\infty}^\infty f(-\overline{w})b(w) \int_{-\infty}^\infty \frac{e^{-i\xi t}d\xi}{(\overline{w}-\xi)^2} du\, dv$$

$$= ct^{-1/2} \int_0^\infty \int_{-\infty}^\infty f(-\overline{w})b(w)te^{-i\overline{w}t}du\, dv$$

(as any book of tables of Fourier transforms will show)

$$= ct^{1/2} \int_0^\infty \int_{-\infty}^\infty f(-\overline{w})b(w)e^{-i\overline{w}t}du\, dv$$

$$= ct^{1/2} \int_u b(w)\overline{\left(\overline{f}(-\overline{w})e^{iwt}\right)}\, dw$$

$$= c\langle x^{-1/2}\hat{b}(x),\ x^{-1/2}(\overline{f}(-u)e^{iut})\hat{}\,(x)\rangle_{\mathcal{L}^2(0,\infty)}$$

(since $\mathcal{F}_1$ is an isometry)

$$= ct^{1/2}\langle x^{-1/2}\hat{b}(x),\ x^{-1/2}\overline{f}(-u)(x-t)\rangle$$

$$= ct^{1/2} \int_0^\infty x^{-1}\hat{b}(x)\hat{f}(x-t)dx = ct^{1/2} \int_0^\infty \frac{1}{s+t}\hat{b}(s+t)\hat{f}(s)ds$$

(by letting $x - t = s$).

On the other hand,

$$(K_b\mathcal{F}_1 f)(t) = \int_0^\infty \frac{\hat{b}(s+t)}{s+t}\sqrt{st}(\mathcal{F}_1 f)(s)ds$$

$$= c\int_0^\infty \frac{\hat{b}(s+t)}{s+t}\sqrt{st}s^{-1/2}\hat{f}(s)ds$$

$$= ct^{1/2}\int_0^\infty \hat{b}(s+t)\frac{1}{s+t}\hat{f}(s)ds.$$

Thus $K_b\mathcal{F}_1 = \mathcal{F}_1 H_b$ giving us the desired unitary equivalence.

Remarks. (1). If $\varphi \in \mathcal{B}(\mathbf{D})$ then a simple argument involving polar coordinates shows that $\varphi \in \mathcal{L}_a^2(\mathbf{D})$. In Proposition 2, we consider only functions which are images of functions of the form $z^2\varphi(z)$ with $\varphi \in \mathcal{B}(\mathbf{D})$.

(2). If $\varphi \in BMOA(\subseteq \mathcal{B}(\mathbf{D}))$ then $S_\varphi \in \mathcal{B}(H^2(T))$ is unitarily equivalent to $S_k \in \mathcal{L}^2(0,\infty)$ i.e., to an integral operator with

$$\text{kernel } k(s,t) = \hat{\psi}(s+t); \psi(t) = \left(\frac{1+it}{1-it}\right)\varphi\left(\frac{1+it}{1-it}\right)$$

and $H_\varphi \in \mathcal{B}(\mathcal{L}_a^2(\mathbf{D}))$ is unitarily equivalent to an integral operator with

$$\text{kernel } = \frac{\sqrt{st}}{s+t}\hat{b}(s+t)$$

with

$$b(w) = \left(\frac{w-i}{w+i}\right)^2 \varphi\left(\frac{w-i}{w+i}\right).$$

(3). There is an integral analogue of the matricial lifting of Hankel matrices to Bergman–Hankel matrices described in [3]. If b is an analytic function on the u.h.p. and $b(x) \in BMO(\mathbb{R})$ such that

$$f \mapsto \int_0^\infty \hat{b}(s+t)f(s)ds$$

defines a bounded operator on $\mathcal{L}^2(0,\infty)$, then so does

$$f \mapsto \int_0^\infty \hat{b}(s+t)\frac{\sqrt{st}}{s+t}f(s)ds.$$

This is a simple consequence of the fact that for a fixed t, the multiplier function $\frac{\sqrt{st}}{s+t}$ peaks to $\frac{1}{2}$ at $s = t$.

(4). A more general version of Proposition 2 is valid for Hankel operators on the Bergman space $A^{s,\gamma}$. They turn out to be unitarily equivalent to integral operators with kernels

$$k(s,t) = \frac{(st)^{\gamma+1/2}}{(s+t)^{2\gamma+1}}\hat{b}(s+t).$$

We are grateful to R. Rochberg for giving us the details of that computation.

Example. If $b(z) = \log z$, $-\pi < Im\ b(z) < \pi$, $b(x) = -i\ sgn\ x$, $\hat{b}(t) = PV\ 1/t$, $(t > 0)$. The integral operator

$$(S_k f)(x) = \int_0^\infty \frac{1}{x+y}f(y)dy$$

is bounded. There are many ways to prove this and express it as a multiplication operator. One of them is given in [6, p. 18]. In terms of $e_m(x) = e^{-x/2}\mathcal{L}_n(x)$, the matrix of S_k is easily found.

$$\langle S_k e_n, e_0 \rangle = \int_0^\infty \int_0^\infty \frac{1}{1+x}\mathcal{L}_n(tx)e^{-t(1+x)/2}dt\ dx.$$

Looking up the Laplace transform of $\mathcal{L}_n$ in a table we now have:

$$\langle S_k e_n, e_0 \rangle = 2\int_0^\infty \frac{(x-1)^n}{(1+x)^{n+2}}\ dx$$

$$= 2\int_0^\infty \frac{(x-1)^n}{(1+x)^{n+2}}\ dx = \frac{1}{n+1}[1-(-1)^{n+1}].$$

40

So the integral operator S_k is unitarily equivalent to the Hankel operator S_b on $H^2(\mathbf{R})$ and the integral operator

$$f \overset{K_b}{\to} \int_0^\infty \frac{\sqrt{xy}}{(x+y)^2} f(y)dy$$

is unitarily equivalent to the Hankel operator H_b on A^2. A standard computation using Mellin transform shows that the integral operator K_b is equivalent to $M_{t\ csch(t)}$ on $\mathcal{L}^2(\mathbf{R})$ and hence has spectrum $[0, 1]$.

References

1. P. Duren, *Theory of H^p Spaces,* Academic Press, New York (1970).

2. John Garnett, Bounded analytic functions, Mathematics (96), Academic Press, New York (1981).

3. P. Ghatage, Lifting Hankel operators from the Hardy space to the Bergman space, to appear in The Rocky Mountain Journal of Mathematics.

4. D. Luecking, Representation and duality in weighted spaces of analytic functions, Indiana University Math. J. **34** (1985), 319–336.

5. S. C. Power, Hankel operators on Hilbert space, Bull. London Math. Soc. **12** (1980), 422–442.

6. S. C. Power, *Hankel Operators on Hilbert Space*, Pitman (1981).

7. R. Rochberg, Decomposition theorems for Bergman spaces and their applications, Operators and Function Theory (Ed. S. C. Power), NATO ASI Series C(153), (1985).

8. R. Rochberg, Trace ideal criteria for Hankel operators and commutators, Indiana University Math. J. **31** (1982), 913–925.

9. K. M. R. Stroethoff, Characterization of the Bloch space and related spaces, Ph.D. Thesis, Michigan State University (1987).

10. G. Szegö, Orthogonal polynomials, AMS Colloquium Publications **23**, (1975).

Pratibha Ghatage

Department of Mathematics

Cleveland State University

Cleveland, Ohio 44115

R. Ji and S. Pedersen

Full group C*-algebras and the generalized Kadison conjecture

The problem we want to discuss here is: Does the full group C^*-algebra $C^*(G)$ of a torsion free discrete group contain a nontrivial projection? Of course, one can ask the same question for the reduced group C^*-algebra $C_r^*(G)$, but we will mainly be interested in $C^*(G)$. Let us begin by giving a few historical remarks.

In the 60's Kaplansky and Dixmier asked: Is every simple unital C^*-algebra generated by projections? Effros and Hahn asked: Does an irrational rotation algebra A_Θ have a proper projection? And Kadison (see [9]) conjectured that $C_r^*(F_2)$ is simple and does not contain a non-trivial projection. In the 70's most of these problems were solved. In particular the following facts were established. Blackadar gave an example of a simple non-unital C^*-algebra without non-zero projections. Powers [9] showed that $C_r^*(F_2)$ is simple and has a unique brace. Cohen [4] pointed out that $C^*(F_2)$ has no non-trivial projections. And Choi [3] showed that $C^*(F_2)$ has a faithful finite trace, has no pure hypo-normal elements and gave a simple proof of Cohen's result. Finally, Powers and Rieffel found that A_Θ contain proper projections. The Kadison conjecture was finally settled by Pimsner and Voiculescu (1981) [8]. They proved that $K_o(C_r^*(F_2)) \simeq \mathbb{Z}[1]$, combining this with the canonical trace yield the Kadison conjecture. Let us also mention that the generalized Kadison conjecture (there is no non-trivial projection in $C_r^*(G)$, if G is torsion free, countable and discrete) is a consequence of the Baum–Connes conjecture [2]. It is known that the Baum–Connes conjecture holds for a large class of groups [2, 7]. It is also known that $C^*(G) \simeq C_r^*(G)$, if G is amenable.

The basic idea behind our approach to the problem studied here is that the non-existence of a proper projection in $C^*(G)$ is due to a certain connectedness of the representations of G.

Example. If G is abelian and torsion free, then $C_r^*(G) \simeq C(\hat{G})$ and $\hat{G}$ (the dual) is connected.

The technique is a generalization of Choi's based on some constructions from [6]. We would like to thank Larry Brown for some very stimulating discussions.

In the following G will be a discrete group and K an infinite dimensional Hilbert space. $D(G,K) = D(G) = D$ is the space of all unitary representations of G on K. $D(G)$ is equipped with the following topology. We say that $\{\pi^t\} \subseteq D(G)$ converge to $\pi \in D(G)$ if $\pi^t(g)$ converge to $\pi(g)$ in norm for each g in G. This determines a Hausdorff topology on $D(G)$. A representation π is *maximal* if it extends to a faithful representation of $C^*(G)$.

Definition. *We say that G has property (MC) (resp. (C)) if some maximal (resp. all) element(s) in $D(G)$ is connected to the trivial representation on K by a path in $D(G)$. Clearly, if G has (C), then it has (MC).*

Definition. *Let $C(D)$ be the set of functions from $D(G)$ into $\mathcal{L}(K)$ that satisfy 1) $f(\pi) \in \pi(C^*(G))$ and 2) if $\ker \pi \supseteq \ker \pi^1$, then $f(\pi) = \tilde{\pi}(f(\pi^1))$. Where $\tilde{\pi} : C^*(G)/\ker \pi^1 \to \mathcal{L}(K)$ is determined by $\tilde{\pi}([a]) = \pi(a)$, for a in $C^*(G)$.*

Proposition 1. *Each function in $C(D)$ is continuous, i.e., if π^t converge to π in $D(G)$ then $f(\pi^t)$ converge to $f(\pi)$ w.r.t. the operator norm topology.*

Theorem 2. *$C(D)$ equipped with the natural algebraic operators and the supremum norm is a C^*-algebra isomorphic to $C^*(G)$.*

Sketch of Proof. The first part is trivial. An isomorphism of $C(D)$ onto $C^*(G)$ is given by the evaluation $f \to f(\pi)$, where π is any (fixed) maximal element of D.

Corollary 3. *If G has property (MC), then $C^*(G)$ has no non-trivial projection.*

Proof. Let P be a non-trivial projection in $C^*(G)$. Let π be a maximal element of $D(G)$ connected to the trivial representation by the path $\pi^t, 0 \le t \le 1$ in $D(G)$, $\pi^1 = \pi$ and $\pi^0 = $ trivial representation. Then $\pi^0(P) = 1$ or 0, we may assume $= 0$. Let f in $C(D)$ be determined by $f(\sigma) = \sigma(P)$. Then $\|f(\pi^t)\|$ is continuous in t and $\|f(\pi^1)\| = 1$ (since $\pi = \pi^1$ is faithful) and $\|f(\pi^0)\| = 0$ contradiction.

Proposition 4. *A countable free abelian group has property (C).*

Proof. Let $\{x_j\}$ be a set of generators for G, and let π be a unitary representation. Then there exist bounded commuting selfadjoint operators A_j, such that $\pi(x_j) = \exp iA_j$. Then the family of representations $\{\pi^t\}$ determined by $\pi^t(x_j) = \exp itA_j, 0 \le t \le 1$, give the wanted path.

Proposition 5. *If G and H has property (MC) and H is amenable then $G \oplus H$ also has (MC).*

44

Proof. Let π and σ be maximal in $D(G)$ and $D(H)$ respectively. Then $\pi \otimes \sigma$ is maximal in $D(G \oplus H)$, this uses that $C^*(G \oplus H) \simeq C^*(G) \otimes_{\min} C^*(H)$, which in turn follow from the assumption that H is amenable. If $\{\pi^t\}$ and $\{\sigma^t\}$ are paths for π and σ. Then $\{\pi^t \otimes \sigma^t\}$ is a path for $\pi \otimes \sigma$.

Proposition 6. If G and H have property (C), then so does the free product $G * H$.

Proof. Since G and H are unrelated in $G * H$, the obvious path works.

Lemma 7. If G has (C) then so does any subgroup of G.

Theorem 8. If G has (C) and H is a subgroup of G, then the amalgamated free product $G *_H G$ has (MC).

Example 9. $G = \mathbb{Z}, H = 2\mathbb{Z}$. Let U and V be unitaries on K, satisfying $U^2 = V^2$. We can assume $C^*(U, V) = C^*(G *_H G)$. Then the two matrices

(1)
$$\begin{pmatrix} U & O \\ O & V \end{pmatrix} \quad \text{and} \quad \begin{pmatrix} V & O \\ O & U \end{pmatrix}$$

determine a maximal representation of $G *_H G$. And the matrices

$$\begin{pmatrix} U & O \\ O & V \end{pmatrix} \quad S_t \begin{pmatrix} V & O \\ O & V \end{pmatrix} S_t^*$$

where

$$S_t = \begin{pmatrix} \cos t & \sin t \\ -\sin t & \cos t \end{pmatrix}$$

determine a path from (1) to

(2)
$$\begin{pmatrix} U & O \\ O & V \end{pmatrix} \quad \text{and} \quad \begin{pmatrix} U & O \\ O & V \end{pmatrix}.$$

If U_t and V_t are paths from U (respectively V) to I, then

$$\begin{pmatrix} U_t & O \\ O & V_t \end{pmatrix} \quad \text{and} \quad \begin{pmatrix} U_v & O \\ O & V_t \end{pmatrix}$$

determine a path from (2) to the trivial representation of $G *_H G$.

The general case, even more general than the statement of Theorem 8 suggest, is motivated by this example.

It can also be proven that 1) if G has property (MC) so does $G \times_\phi \mathbb{Z}$, if ϕ is periodic, and 2) if G and H are amenable (MC)-groups so is the restricted wreath product of G and H.

The fact that $C^*(G)$ for groups with property (MC) do not have proper projections, stem essentially from the fact that the path $\{\pi^t\}$ $0 \leq t \leq 1$, from a maximal representation π^1 to the trivial representation π^o, lead to a continuous field of C^*-algebras $\pi^t(C^*(G))$, where one $(\pi^1(C^*(G)))$ is $C^*(G)$ and one $(\pi^o(C^*(G)))$ does not contain any proper projections. We have not been able to decide whether or not torsion free nilpotent groups have property (MC). But using the above idea, we can show that the full group C^*-algebra of a torsion free nilpotent group does not contain any proper projection, specifically we have

Theorem 10. *Let $O \to C \to G \to N \to O$ be a central extension of groups. If N is amenable, then $C^*(G)$ is a continuous field of C^*-algebras over the dual of C.*

This theorem is probably known to some people, e.g. , the authors of [1, 5, 10] but we have not been able to locate a complete proof. As an immediate corollary we get that, if furthermore C is torsion free and $C^*(N)$ has no proper projections, then neither do $C^*(G)$. An induction now proves that the full group C^*-algebra of a torsion free nilpotent group is without proper projections.

Example. Let G be the universal group generated by U, V and the relation $VU = UV^2$. Then G is torsion free, but it does not have property (MC).

Proof. Let U_t and V_t be a continuous family of unitaries satisfying $V_t U_t = U_t V_t^2, 0 \leq t \leq 1, V_o = U_o = I$. Then

$$(3) \qquad\qquad sp(V_t) \subseteq \left\{ e^{is} \mid -\delta \leq s \leq \delta \right\}$$

for arbitrarily small δ, provided $t = t(\delta)$ is small enough. But V_t is unitarily equivalent to V_t^2, hence $sp(V_t) = sp(V_t^2)$, contradicting (3) unless $V_t = I$. We conclude $V_t = I$ for all t.

Actually, any group of Kazdan's property T does not have property (MC). Details will appear elsewhere.

References

1. T. Anderson and W. Paschke, The rotation algebra, MSRI, 1985, preprint.

2. P. Baum and A. Connes, Geometric K-theory for lie groups and foliations, IHES, preprint.

3. M.–D. Choi, The full C^*-algebra of the free group on two generators, Pac. J. Math. **87** (1980), 41–48.

4. T. Cohen, C^*-algebras without idempotents, J. Funct. Anal. **33** (1979), 211–216.

5. G. Elliott, On the K-theory of the C^*-algebra generated by a projective representation of a torsion free discrete abelian group, *Operator Algebra and Group Representations*, Pitman, London, 1983, 157–184.

6. P. Kruszynski and S. Woronowicz, A non-commutative Gelfand–Naimark theorem, J. Operator Theory **8** (1982), 361–381.

7. M. Pimsner, KK-groups of crossed products by groups acting on trees. Inv. Math. **86** (1986), 603–634.

8. M. Pimsner and D. Voiculescu, K-groups of reduced crossed products by free groups, J. Operator Theory **8** (1982), 131–156.

9. R. Powers, Simplicity of the C^*-algebra associated with the free group on two generators, Duke Math. J. **42** (1975), 151–156.

10. M. A. Reiffel, Induced representations of C^*-algebras, Adv. in Math. **13** (1974), 176–257.

Ronghui Ji
Department of Mathematics
Indiana University–Purdue University at Indianapolis
Indianapolis, IN 46223

and

Steen Pedersen
Department of Mathematics
Wright State University
Dayton, OH 45435

Partially supported by NSF, the first named author is grateful to this support.

Gert K. Pedersen

The corona construction

§0. Introduction

These notes cover an expanded version of a series of lectures delivered at the Great Plains Operator Theory Seminar held in May 1988 in Indianapolis. A spirited audience and a faultless organization, run by Bernard Morrel, made the event fully comparable to that other great race in the city, held a few weeks later – at least to the mathematical world.

Apart from standard facts from C^*-algebra theory, the corona construction and its applications use only a few very basic notions, centered around approximate identities and multiplier theory. Thus it has been possible to write a fairly self-contained account and, hopefully, to stress motivation and exemplification more than could be allowed in the original papers. The contents is organized as follows.

1. Preliminaries, p. 49
2. Approximate Units, p. 51
3. The Multiplier Algebra, p. 53
4. Properties of Multipliers, p. 56
5. The Corona Algebra, p. 59
6. Properties of Corona Algebras, p. 61
7. *SAW** Algebras, p. 66
8. Kasparovs Technical Theorems, p. 71
9. Weak Polar Decomposition, p. 73
10. Derivations and Morphisms, p. 78
11. A Lifting Problem, p. 85
    References, p. 90

§1. Preliminaries

By a C^*-*algebra* we mean a complex Banach algebra $\mathfrak{A}$ with an involution, $*$, that satisfies the C^*-condition: $\| A^*A \|=\| A \|^2$, for every A in $\mathfrak{A}$. A *representation* of $\mathfrak{A}$ is a pair $(\rho, \mathfrak{H})$, where $\mathfrak{H}$ is a complex Hilbert space, and ρ is a *-preserving homomorphism of $\mathfrak{A}$ into the C^*-algebra $\mathbb{B}(\mathfrak{H})$ consisting of the bounded linear operators on $\mathfrak{H}$. The representation is *non-degenerate*, if $\rho(A)x = 0$ for some x in $\mathfrak{H}$ and every A in $\mathfrak{A}$ implies that $x = 0$, i.e. if the subspace

$$\mathfrak{H}_0 = \{x \in \mathfrak{H} \mid \rho(\mathfrak{A})x = 0\}$$

is zero. Since in any case both $\mathfrak{H}_0$ and $\mathfrak{H}_0^\perp$ are invariant subspaces for $\rho(\mathfrak{A})$ (as $\rho(\mathfrak{A})$ is a *-algebra), we can always pass from a non-zero representation $(\rho, \mathfrak{H})$ to the non-degenerate representation $(\rho | \mathfrak{H}_0^\perp, \mathfrak{H}_0^\perp)$. A representation $(\rho, \mathfrak{H})$ of $\mathfrak{A}$ is *faithful* if ρ is injective, i.e. $\ker \rho = 0$. It follows from spectral theory that a faithful representation is an isometry, and from that it is easy to show that every representation is norm-decreasing with a closed image. Thus $\rho(\mathfrak{A})$ is a C^*-algebra of operators, isomorphic to the quotient algebra $\mathfrak{A} / \ker \rho$.

Two representations $(\rho_1, \mathfrak{H}_1)$ and $(\rho_2, \mathfrak{H}_2)$ of a C^*-algebra $\mathfrak{A}$ are *unitarily equivalent* if there is a unitary operator U of $\mathfrak{H}_1$ onto $\mathfrak{H}_2$, such that $U \rho_1(A)U^* = \rho_2(A)$ for every A in $\mathfrak{A}$. A non-degenerate representation $(\rho_u, \mathfrak{H}_u)$ is *universal* if every other representation of $\mathfrak{A}$ is unitarily equivalent to a subrepresentation of $(\rho_u, \mathfrak{H}_u)$, i.e. a representation of the form $\rho_u | \mathfrak{H}, \mathfrak{H})$ for some invariant subspace $\mathfrak{H}$ of $\mathfrak{H}_u$. The existence of a faithful, universal representation of any given abstract C^*-algebra is a celebrated result of Gelfand and Naimark. The uniqueness, up to unitary equivalence, of the universal representation is inherent from its definition. From the construction of the universal representation of $\mathfrak{A}$ it follows that every bounded linear functional φ of $\mathfrak{A}$ has the form

$$\varphi(A) = (\rho_u(A)x|y), \quad A \in \mathfrak{A},$$

for some vectors x and y in $\mathfrak{H}_u$; and that we can choose $x = y$ if φ is *positive*, i.e. $\varphi(A^*A) \geq 0$, $A \in \mathfrak{A}$, see [20, 3.7-3.8].

The following three examples of C^*-algebras will be used repeatedly in the following.

Example 1.1. Let $C_0(\mathbb{R}^n)$ denote the algebra of continuous (complex) functions on $\mathbb{R}^n$, $n \in \mathbb{N}$, vanishing at infinity. More generally, consider $\mathfrak{A} = C_0(X)$, where X is a locally compact Hausdorff space. These algebras are all commutative and, as shown by Gelfand, they are the most general examples. Thus, if $\mathfrak{A}$ is a commutative C^*-algebra, then with X as the locally compact space of characters on $\mathfrak{A}$ we have an isometric *-isomorphism between $\mathfrak{A}$ and $C_0(X)$.

50

Example 1.2. Let $\mathfrak{K}$ denote the C^*-algebra of compact operators on the Hilbert space ℓ^2. Thus $\mathfrak{K}$ consists of the operators T in $\mathsf{B}(\ell^2)$ for which the map $T : \ell^2 \to \ell^2$ is weak-norm continuous on bounded subsets. Equivalently, $\mathfrak{K}$ is the norm closure of the algebra of operators in $\mathsf{B}(\ell^2)$ of finite rank, or, more bluntly, the closure of the algebra of operators with finite matrices (in the standard basis for ℓ^2). The algebra $\mathfrak{K}$ is the only infinite-dimensional, separable C^*-algebra with a single irreducible representation (up to unitary equivalence), and the given representation on ℓ^2 is very nearly equivalent to the universal representation. In fact, the universal representations of $\mathfrak{K}$ is obtained by adding a countable number of copies of the given representation.

Example 1.3. Take $\mathfrak{A} = C_0(\mathsf{R}^n, \mathfrak{K})$, the algebra of $\mathfrak{K}$-valued continuous functions on R^n, vanishing at infinity, equipped with pointwise sum, product and involution and with supremum norm. Whereas $C_0(\mathsf{R}^n)$ is the typical example of a commutative C^*-algebra, and $\mathfrak{K}$ is the simplest example of a very non-communicative, infinite-dimensional C^*-algebra (with no non-trivial, closed ideals), the algebra $\mathfrak{A}$ is the simplest way of mixing the two prototypes. In fact, $\mathfrak{A}$ is the C^*-completion of the algebraic tensor of the two previous algebras, and is usually written as $\mathfrak{A} = C_0(\mathsf{R}^n) \otimes \mathfrak{K}$ (with an appropriate notion of C^*-tensor product).

§2. Approximate Units

An *approximate unit* for a C^*-algebra $\mathfrak{A}$ is a net $(E_\lambda)_{\lambda \in \Lambda}$ in $\mathfrak{A}$, such that $E_\lambda \geq 0$ and $\| E_\lambda \| \leq 1$ for all λ, and such that

$$(*) \qquad \lambda < \mu \Rightarrow E_\lambda \leq E_\mu \quad \text{(in the order from } \mathfrak{A}_+),$$

$$(**) \qquad \| A - E_\lambda A \| \to 0 \ \text{(and} \ \| A - E_\lambda A \| \to 0) \quad \forall A \in \mathfrak{A}.$$

As shown by Segal, every C^*-algebra has an approximate unit; in fact, as Dixmier observed, one may use the net $\{ T \in \mathfrak{A}_+ \mid \| T \| < 1 \}$ with the induced order. In order to utilize arguments involving differences of consecutive elements we need *countable* approximate units. Such ones will exist if and only if $\mathfrak{A}$ has a *strictly positive element* H; i.e. an element H such that $(Hx|x) > 0$ for every non-zero vector x in $\mathfrak{H}_u$ (equivalently $\varphi(H) > 0$ for every state φ of $\mathfrak{A}$). Algebras with this property are called *σ-unital*, and from now on all (non-unital) C^*-algebras will be assumed to be σ-unital. Since

51

every separable C^*-algebra is σ-unital, this is not a serious limitation. To explain the restriction, take $\mathfrak{A} = C_0(X)$. It is a fact that $\mathfrak{A}$ is σ-unital if and only if X is σ-compact. Furthermore, a locally compact, σ-compact Hausdorff space is normal and para-compact; properties that may fail in the absense of σ-compactness. In the authors opinion, being σ-unital is a basic property for a C^*-algebra, second in importance only to separability.

Let H be a strictly positive element in a σ-unital C^*-algebra $\mathfrak{A}$. Then $\mathfrak{A}H$ (and $H\mathfrak{A}$) is dense in $\mathfrak{A}$ – otherwise its closure would be a proper left ideal, and therefore be annihilated by a (pure) state of $\mathfrak{A}$. From this observation we can easily deduce the existence of an approximate unit. Taking $\| H \| \leq 1$, we let $E_n = H^{1/n}$ to obtain a countable, *commutative* approximate unit for $\mathfrak{A}$. Since the same argument will work with $E_n = f_n(H)$, provided that $f_n \in C([0,1])$ with $f_n(0) = 0$ and $f_n(x) \nearrow 1$, $\forall x \neq 0$, we can do better: Let f_n be the piecewise linear function on $[0,1]$ such that $f(x) = 0$ for $x \leq (n+1)^{-1}$ and $f(x) = 1$ for $x \geq n^{-1}$. Then with $E_n = f_n(H)$ we have an approximate unit where $E_{n+1}E_n = E_n$ for all n. In particular, the differences $D_n = E_n - E_{n-1}$ will be pairwise orthogonal, $D_n D_m = 0$, if $|n - m| \geq 2$.

An approximate unit for $\mathfrak{A}$ is *central* if $E_n A = A E_n$ for every n and every A in $\mathfrak{A}$. If $\mathfrak{A} \subset B(\mathfrak{H})$ and $\mathfrak{M}$ is another subset of $B(\mathfrak{H})$ we say that an approximate unit $(E_\lambda)_{\lambda \in \Lambda}$ for $\mathfrak{A}$ is *quasi-central with respect to* $\mathfrak{M}$ if

$$\| E_\lambda T - T E_\lambda \| \to 0 \quad \forall T \in \mathfrak{M}.$$

Clearly (E_λ) is quasi-central w.r.t. $\mathfrak{A}$, but much more is true.

Theorem 2.1. *Every C^*-algebra $\mathfrak{A}$, contained as an ideal in a larger C^*-algebra $\mathfrak{M}$, has an approximate unit $(E_\lambda)_{\lambda \in \Lambda}$ which is quasi-central w.r.t. $\mathfrak{M}$. If $(F_\mu)_{\mu \in M}$ is any given approximate unit for $\mathfrak{A}$, we may choose (E_λ) in the convex hull of the set $\{F_\mu \mid \mu \in M\}$.*

Sketch of Proof: Assume that $\mathfrak{A} \subset \mathfrak{M} \subset B(\mathfrak{H}_u)$. Although that can only be ascertained when $\mathfrak{A}$ is an essential ideal in $\mathfrak{M}$ (more about this later), the restriction is easily circumvented by passing to $\mathfrak{M}/\mathfrak{A}^\perp$. Given an approximate unit (F_μ), note that (F_μ) converges strongly to I in $B(\mathfrak{H}_u)$. Thus $F_\mu T - T F_\mu \to 0$ strongly, hence weakly in $B(\mathfrak{H}_u)$. Since $F_\mu T - T F_\mu \in \mathfrak{A}$ when $T \in \mathfrak{M}$, and since every element in $\mathfrak{A}^*$ can be represented as a vector functional on $B(\mathfrak{H}_u)$, we see that $F_\mu T - T F_\mu \to 0$ in the $\sigma(\mathfrak{A}, \mathfrak{A}^*)$-topology on $\mathfrak{A}$. By Hahn-Banach's theorem $\mathrm{conv}\{F_\mu T - T F_\mu\}$ must therefore contain O as a limit point in norm.

$\square$

In the sequel we shall often apply the following version of the result above.

Theorem 2.2. *If $\mathfrak{A}$ is a σ-unital C^*-algebra, contained as in ideal in a C^*-algebra $\mathfrak{M}$, and if (T_k) is a sequence in $\mathfrak{M}$, there is an approximate unit (E_n) for $\mathfrak{A}$ such that*

$$(*) \qquad\qquad E_{n+1} E_n = E_n \quad \text{for all } n$$

$$(**) \qquad\qquad \| E_n T_k - T_k E_n \| < \varepsilon_n \quad \text{for all } k \leq n,$$

where (ε_n) is any prearranged sequence in $\mathbb{R}_+$.

§3. The Multiplier Algebra

Recall that an ideal $\mathfrak{A}$ is an algebra $\mathfrak{M}$ is *essential* if $\mathfrak{J} \cap \mathfrak{A} \neq 0$ for every non-zero ideal $\mathfrak{J}$ of $\mathfrak{M}$. Equivalently, the annihilator

$$\mathfrak{A}^{\perp} = \{ T \in \mathfrak{M} \mid T\mathfrak{A} = 0 \}$$

is the zero ideal.

Define a *unitization* of a C^*-algebra $\mathfrak{A}$ to be an embedding of $\mathfrak{A}$ as an essential ideal of a unital C^*-algebra $\mathfrak{M}$. It is easy to show that, when $\mathfrak{A} = C_0(X)$, there is a one-to-one correspondence between unitizations of $\mathfrak{A}$ and compactifications of X. Thus unitizations may be regarded as the non-commutative analogues of compactifications. In the commutative case, $\mathfrak{A} = C_0(X)$, it is well known that there is a minimal compactification, $X \cup \{\infty\}$, and a maximal compactification $\beta(X)$ – the Stone-Čech compactification. Obviously the minimal compactification has an analogue in the non-commutative case, viz. the minimal unitization $\mathfrak{A} \oplus \mathbb{C}$. Less obvious is the fact that also the Stone-Čech compactification generalizes:

Theorem 3.1. *Every C^*-algebra $\mathfrak{A}$ has a (unique) maximal unitization $M(\mathfrak{A})$, called the multiplier algebra. Thus if $\iota : \mathfrak{A} \to \mathfrak{M}$ is a unitization of $\mathfrak{A}$, there is a natural embedding of $\mathfrak{M}$ as a C^*-subalgebra of $M(\mathfrak{A})$, i.e. $\mathfrak{A} \subset \mathfrak{M} \subset M(\mathfrak{A})$.*

To explain the construction of $M(\mathfrak{A})$ abstractly, define a *left (right) centralizer* of $\mathfrak{A}$ to be a linear map $L : \mathfrak{A} \to \mathfrak{A}$ ($R : \mathfrak{A} \to \mathfrak{A}$) satisfying $L(AB) = L(A)B$ ($R(AB) = AR(B)$) for all pairs A, B in $\mathfrak{A}$. Using factorization theorems it can be shown that centralizers are automatically continuous, cf. [20, 3.12.2]. Define an involution on centralizers as usual:

$$L^*(A) = L(A^*)^* \ , \quad R^*(A) = R(A^*)^*,$$

53

and note that it intertwines left and right. A *double centralizer* is a pair (R, L) of centralizers such that $R(A)B = AL(B)$ for all A, B in $\mathfrak{A}$. On the set $\tilde{M}(\mathfrak{A})$ of double centralizers, which is a Banach space under the norm

$$\| (R, L) \| = \| R \| \quad (= \| L \|),$$

we define involution as

$$(R, L)^* = (L^*, R^*),$$

and we define product as

$$(R_1, L_1)(R_2, L_2) = (R_2 R_1, L_1 L_2).$$

Using the C^*-norm property on $\mathfrak{A}$, it follows that $\tilde{M}(\mathfrak{A})$ is a C^*-algebra. Now define an embedding of $\mathfrak{A}$ into $\tilde{M}(\mathfrak{A})$ by mapping each element A into the pair (R_A, L_A) in $\tilde{M}(\mathfrak{A})$ defined by

$$R_A(B) = BA, \quad L_A(B) = AB, \quad B \in \mathfrak{A}.$$

Since by computation

$$(R, L)(R_A, L_A) = (R_{L(A)}, L_{L(A)}),$$
$$(R_A, L_A)(R, L) = (R_{R(A)}, L_{R(A)}),$$

it follows that $\mathfrak{A}$ is an (essential) ideal in $\tilde{M}(\mathfrak{A})$.

A more concrete realization of the multiplier algebra is also possible. Consider $\mathfrak{A}$ as a subalgebra of $\mathsf{B}(\mathfrak{H})$ for some Hilbert space $\mathfrak{H}$ (cf. §1). Now define

$$M(\mathfrak{A}) = \{T \in \mathsf{B}(\mathfrak{H}) \mid T\mathfrak{A} + \mathfrak{A}T \subset \mathfrak{A}\},$$

i.e., take the idealizer of $\mathfrak{A}$ in $\mathsf{B}(\mathfrak{H})$. It is clear that if $\mathfrak{A}$ acts non-degenerately on $\mathfrak{H}$, then $\mathfrak{A}$ is an essential ideal in $M(\mathfrak{A})$. Moreover, if $T \in M(\mathfrak{A})$, then the pair (R_T, L_T) defined as above gives a double centralizer. The important thing is that the converse also holds. If namely $(R, L) \in \tilde{M}(\mathfrak{A})$, let (E_λ) be an approximate unit for $\mathfrak{A}$ and choose weak limit points S and T for the bounded nets $(R(E_\lambda))$ and $(L(E_\lambda))$, respectively. Passing if necessary to a subnet we then have, for all A, B in $\mathfrak{A}$,

$$R(A) = \lim R(AE_\lambda) = \lim AR(E_\lambda) = AS,$$

and similarly $L(B) = TB$. Moreover,

$$ASB = R(A)B = AL(B) = ATB,$$

so from the non-degeneracy of $\mathfrak{A}$ it follows that $S = T$. We see from this that S and T are the unique limit points of $R(E_\lambda)$ and $L(E_\lambda)$, and that $S(= T)$ is an element of $M(\mathfrak{A})$ such that $(R, L) = (R_S, L_S)$. Consequently we may identify $M(\mathfrak{A})$ and $\tilde{M}(\mathfrak{A})$. Combining the abstract and the concrete realization of the multiplier algebra it is easy to verify Theorem 3.1.

Now consider $\mathfrak{A}$ in its universal representation (cf§1), and take the weak closure of $\mathfrak{A}$ in $\mathsf{B}(\mathfrak{H}_u)$. This *enveloping von Neumann algebra* is isometrically *-isomorphic to $\mathfrak{A}^{**}$ – the second dual of $\mathfrak{A}$ – cf. [20, 3.7.8] and from the previous considerations we have the inclusions

$$\mathfrak{A} \subset M(\mathfrak{A}) \subset \mathfrak{A}^{**} \subset \mathsf{B}(\mathfrak{H}_u).$$

Given a subset $\mathfrak{M}$ in $\mathsf{B}(\mathfrak{H}_u)_{sa}$, we let $\mathfrak{M}^m$ denote the set of (self-adjoint) operators T in $\mathsf{B}(\mathfrak{H}_u)$ for which there is an increasing net (T_λ) in $\mathfrak{M}$ with T as its least upper bound. Thus $(Tx|x) = \lim(T_\lambda x|x)$ for each x in $\mathfrak{H}_u$, from which one may deduce that actually $T_\lambda \to T$, strongly. Our notation for this phenomenon will be $T_\lambda \nearrow T$. Similarly we define $\mathfrak{M}_m$ as the set of limit points of monotone decreasing nets from $\mathfrak{M}$. Thus $\mathfrak{M}_m = -(-\mathfrak{M})^m)$. With this notation we can formulate an order-theoretic characterization of (self-adjoint) multipliers:

Theorem 3.2.

$$M(\mathfrak{A})_{sa} = (\tilde{\mathfrak{A}}_{sa})^m \cap (\tilde{\mathfrak{A}}_{sa})_m,$$

where $\tilde{\mathfrak{A}}$ denotes the minimal unitization $\mathfrak{A} \oplus \mathbb{C}1$ in $\mathsf{B}(\mathfrak{H}_u)$.

Proof: Apply Dini's lemma, see [20, 3.12.9].

$\square$

To understand the content of the theorem, consider the commutative case, cf. Example 1.1. If $\mathfrak{A} = C_0(X)$ let $\mathfrak{F}_b(X)$ denote the von Neumann algebra (on $\ell^2(X)$) of all bounded functions on X. Although this is not the universal representation of $\mathfrak{A}$ (but the reduced atomic representation), it is large enough for our purposes. It is easy to verify that $(\tilde{\mathfrak{A}}_{sa})^m$ $(= (C(X \cup \{\infty\})_{sa})^m)$ is the set of bounded, real-valued, lower semicontinuous functions on X. Thus $(\tilde{\mathfrak{A}}_{sa})^m \cap (\tilde{\mathfrak{A}}_{sa})_m$ is the set of real-valued, bounded functions on X that are both upper and lower semicontinuous, i.e. it coincides with the algebra $C_b(X)_{sa}$, which, of course, is the self-adjoint part of the idealizer of $C_0(X)$ in $\mathfrak{F}_b(X)$.

In order to give also a topological characterization of the multiplier algebra, we define the *strict topology* on $M(\mathfrak{A})$ (or on $\mathsf{B}(\mathfrak{H}_u)$) as the weak topology induced by the seminorms

$$T \to \| \, TA \, \| + \| \, AT \, \|, \quad A \in \mathfrak{A},$$

see [8]. In this topology $\mathfrak{A}$ is a topological algebra with a continuous involution, and the completion of $\mathfrak{A}$ (equivalently: the strict closure of $\mathfrak{A}$ in $\mathsf{B}(\mathfrak{H}_u)$) is precisely $M(\mathfrak{A})$. Note that if $\mathfrak{A}$ is σ-unital, the strict topology is metrizable on bounded subsets. In fact, if H is a strictly positive element, let

$$d(S,T) = \| \, (S-T)H \, \| + \| \, H(S-T) \, \|$$

for S, T in $M(\mathfrak{A})$ (or in $\mathsf{B}(\mathfrak{H}_u)$).

Let us apply the previous notions to the three examples mentioned in §1. If $\mathfrak{A} = C_0(\mathbf{R}^n)$ we have $M(\mathfrak{A}) = C_b(\mathbf{R}^n)$ $(= C(\beta \mathbf{R}^n))$ as mentioned above. If we use only the seminorms $T \to \| \, AT \, \|$, where $A \in C_c(\mathbf{R}^n)$, then we obtain the topology of uniform convergence on compact subsets. The strict topology on $C_b(\mathbf{R}^n)$ is therefore slightly stronger (since we can now take A in $C_0(\mathbf{R}^n)$), but the two topologies agree on bounded subsets of functions on $\mathbf{R}^n$.

Considering Example 1.2, $\mathfrak{A} = \mathfrak{K}$, we have $M(\mathfrak{K}) = \mathsf{B}(\ell^2)$, so in this case the multiplier algebra coincides with the enveloping von Neumann algebra. If we use only the seminorms $T \to \| \, AT \, \| + \| \, TA \, \|$, where A has finite rank on ℓ^2, we obtain the strong* topology on $\mathsf{B}(\ell^2)$. The strict topology is therefore slightly stronger (though incomparable with the σ-strong *-topology), but the topologies agree on bounded subsets of $\mathsf{B}(\ell^2)$.

In Example 1.3 we have $\mathfrak{A} = C_0(\mathbf{R}^n, \mathfrak{K})$, and we get $M(\mathfrak{A}) = C_b(\mathbf{R}^n, \mathsf{B}(\ell^2)_{s^*})$, the bounded operator-valued functions on $\mathbf{R}^n$ that are continuous from $\mathbf{R}^n$ to $\mathsf{B}(\ell^2)$ with the strong* topology. On bounded subsets of $M(\mathfrak{A})$, the strict topology will be pointwise strong* convergence of functions, uniformly on compact subsets of $\mathbf{R}^n$, see [3, 3.3].

§4. Properties of Multipliers

It is often desirable to be able to write down multipliers with given asymptotical properties. The following is the strongest known construction of this kind. We shall use it repeatedly as a means to convert an approximate property (given by a sequence of elements) to an exact formula (valid modulo $\mathfrak{A}$).

Theorem 4.1. *Let (E_n) be an approximate unit for a (non-unital, but σ-unital) C^*-algebra $\mathfrak{A}$, and let (T_n) be a bounded sequence in $M(\mathfrak{A})$. Then the element*

$$T = \sum_{n=0}^{\infty} (E_n - E_{n-1})^{1/2} T_n (E_n - E_{n-1})^{1/2}$$

belongs to $M(\mathfrak{A})$, the sum being strictly convergent.

Proof: It suffices to consider the case where $0 \leq T_n \leq I$ for all n. To ease the notation we take $E_0 = 0$ and put $F_n = (E_n - E_{n-1})^{1/2}$.

Considering $\mathfrak{A}$ in its universal reprsentation it is easy to see that T exists in $\mathfrak{A}^{**}$ as the limit of a strongly convergent sum. After all, each summand is positive and dominated by F_n^2, and $\sum F_n^2 = I$. However, since $F_n T_n F_n \in \mathfrak{A}_+$ for each n, we see that $T \in (\mathfrak{A}_+)^m$ in the notation from §3. But also

$$I - T = \sum F_n (I - T_n) F_n \in (\mathfrak{A}_+)^m,$$

since $\sum F_n^2 = I$, and thus

$$T \in I - (\mathfrak{A}_+)^m = I + (-\mathfrak{A}_+)_m \subset (\tilde{\mathfrak{A}}_{sa})_m.$$

It follows from Theorem 3.2 that

$$T \in (\mathfrak{A}_+)^m \cap (\tilde{\mathfrak{A}}_{sa})_m \subset M(\mathfrak{A})_{sa}.$$

The alternative proof of the theorem consists of showing that the sum is strictly convergent and thus belongs to $M(\mathfrak{A})$. For this one uses the easy estimate

$$\left\| \left(\sum_{n>m} F_n T_n F_n \right) A \right\| \leq (\sup \| T_n \|) \, \| A(I - E_m)A \|^{1/2},$$

valid for every A in $\mathfrak{A}$.

$\square$

Lemma 4.2. *If $\rho : \mathfrak{A} \to \mathfrak{B}$ is a surjective $*$-homomorphism between C^*-algebras, and if $S, T \in \mathfrak{A}_{sa}$ with $S \leq T$, then for each element B in $\mathfrak{B}_{sa}$ such that $\rho(S) \leq B \leq \rho(T)$ there is an element A in $\mathfrak{A}_{sa}$ with $\rho(A) = B$, such that $S \leq A \leq T$.*

Proof: See [20, 1.5.10] or [9].

$\square$

Theorem 4.3. *For each surjective *-homomorphism $\rho : \mathfrak{A} \to \mathfrak{B}$ between σ-unital C^*-algebras, there is a (unique) surjective extension $\rho^{**} : M(\mathfrak{A}) \to M(\mathfrak{B})$.*

Proof: By Banach space theory we have the injective dual map $\rho^* : \mathfrak{B}^* \to \mathfrak{A}^*$ and the surjective double dual map $\rho^{**} : \mathfrak{A}^{**} \to \mathfrak{B}^{**}$. Realizing $\mathfrak{A}^{**}$ and $\mathfrak{B}^{**}$ as the enveloping von Neumann algebras of $\mathfrak{A}$ and $\mathfrak{B}$, respectively, it is not hard to show that ρ^{**} is a (weakly continuous and normal) *-homomorphism. Since $M(\mathfrak{A})$ and $M(\mathfrak{B})$ are the idealizers of $\mathfrak{A}$ and $\mathfrak{B}$ in $\mathfrak{A}^{**}$ and $\mathfrak{B}^{**}$, respectively, it follows that $\rho^{**}(M(\mathfrak{A})) \subset M(\mathfrak{B})$.

To show the surjectivity of the map, take B in $M(\mathfrak{B})_{sa}$ and choose a strictly positive element H in $\mathfrak{A}_+$. By Theorem 3.2 there are monotone nets (B_λ^+) and (B_λ^-) in $\tilde{\mathfrak{B}}_{sa}$, the first increasing the other decreasing, such that $B_\lambda^+ \nearrow B$ and $B_\lambda^- \searrow B$. From Dini's lemma it follows that the net $\rho(H)(B_\lambda^- - B_\lambda^+)\rho(H)$ in $\mathfrak{B}_+$ decreases uniformly to 0. We can therefore choose sequences (B_n^+) and (B_n^-), the first increasing the other decreasing, such that

$$\| \, \rho(H)(B_n^- - B_n^+)\rho(H) \, \| < n^{-1}$$

for all n. We claim that there are sequences (A_n^+) and (A_n^-) in $\tilde{\mathfrak{A}}_{sa}$, the first increasing the other decreasing, such that $\rho(A_n^+) = B_n^+$, $\rho(A_n^-) = B_n^-$, $A_n^+ \leq A_n^-$, and

$$(*) \qquad\qquad\qquad \| \, H(A_n^- - A_n^+)H \, \| < n^{-1}$$

for all n. Suppose that the claim has been established for all $k \leq n$. By Lemma 4.2 there is then an element A_{n+1}^- in $\tilde{\mathfrak{A}}_{sa}$ such that $\rho(A_{n+1}^-) = B_{n+1}^-$ and $A_n^+ \leq A_{n+1}^- \leq A_n^-$. Applying the lemma again, this time to the pair A_n^+, A_{n+1}^-, we find an element A_{n+1} with $\rho(A_{n+1}) = B_{n+1}^+$ and $A_n^+ \leq A_{n+1} \leq A_{n+1}^-$. Now choose an approximate unit (E_λ) for $\ker \rho$ and consider the net with elements

$$A_\lambda = A_{n+1} + \left(A_{n+1}^- - A_{n+1}\right)^{1/2} E_\lambda \left(A_{n+1}^- - A_{n+1}\right)^{1/2}.$$

Note that $A_{n+1} \leq A_\lambda \leq A_{n+1}^-$ and $\rho(A_\lambda) = B_{n+1}^+$ for all λ. Moreover,

$$H \left(A_{n+1}^- - A_\lambda\right) H = H \left(A_{n+1}^- - A_{n+1}\right)^{1/2} (I - E_\lambda) \left(A_{n+1}^- - A_{n+1}\right)^{1/2} H,$$

from which it follows that

$$\lim_\lambda \|H(A_{n+1}^- - A_\lambda)H\| = \|\rho(H)\left(B_{n+1}^- - B_{n+1}^+\right)\rho(H)\| < (n+1)^{-1}.$$

We can therefore take A_{n+1}^+ to be A_λ for λ sufficiently large, such that

$$\| \, H(A_{n+1}^- - A_{n+1}^+)H \, \| < (n+1)^{-1},$$

which establishes the claim by induction. If A^+ and A^- denotes the limits of the monotone sequences (A_n^+) and (A_n^-), then clearly $A^+ \le A^-$ and $H(A^- - A^+)H = 0$. Since H is strictly positive, this implies that $A^- - A^+ = 0$, so that

$$A^+ = A^- \in (\tilde{\mathfrak{A}}_{sa})^m \cap (\tilde{\mathfrak{A}}_{sa})_m = M(\mathfrak{A})_{sa}.$$

Evidently $\rho^{**}(A^+) = B$, and the surjectivity is established.

$\square$

Remark 4.4. The previous theorem may be regarded as the non-commutative analogue of the Tietze extension theorem. Indeed, if $\mathfrak{A} = C_0(X)$ for some locally compact, σ-compact Hausdorff space X, then the only surjective quotient maps are obtained by taking $\rho(A) = A|Y$, for some closed subset Y of X. As $M(\mathfrak{A}) = C_b(X)$ and $M(\rho(\mathfrak{A})) = C_b(Y)$, we learn from Theorem 4.3 that each bounded continuous function B on Y is the image of (i.e., has an extension to) a bounded continuous function A on X; and that is precisely Tietze's theorem. Note that the σ-unital condition is crucial for the theorem. For only if X is σ-compact can we conclude that local compactness implies normality of the topological space X.

§5. The Corona Algebra

Let $\mathfrak{A}$ be a σ-unital C^*-algebra with multiplier algebra $M(\mathfrak{A})$. We define the *corona algebra* for $\mathfrak{A}$ as the C^*-algebra

$$C(\mathfrak{A}) = M(\mathfrak{A})/\mathfrak{A}.$$

Going back to the examples in §1, we have in 1.2 that $C(\mathfrak{K}) = \mathsf{B}(\mathfrak{H})/\mathfrak{K}$ – the Calkin algebra. This example certainly is the first motivation for studying corona C^*-algebras, to the extent that one is tempted to consider corona algebras as generalized Calkin algebras, which is, maybe, not quite the truth. In Example 1.1 we have

$$C(C_0(\mathbf{R}^n)) = C_b(\mathbf{R}^n)/C_0(\mathbf{R}^n)$$
$$= C(\beta\mathbf{R}^n)/C_0(\mathbf{R}^n) = C(\beta\mathbf{R}^n \setminus \mathbf{R}^n).$$

The name corona algebra originates from this case. For if X is a locally compact Hausdorff space, e.g. $\mathbf{R}^2$, then $\beta X \setminus X$ is called the corona of X in [12]. Such spaces have amused the topologists for many years; witness the papers by W. G. Bade, G. Choquet,

L. Gillman, K. Grove and G. K. Pedersen, M. Henriksen, M. Jerison, R. R. Smith and D. P. Williams. To see how far a corona algebra can stray from the Calkin algebra, the reader should know that the corona of $\mathbf{R}^2$ is a connected, compact space of covering dimension 2 with no convergent sequences and no curves. Thus the corona algebra $C(\beta\mathbf{R}^2 \setminus \mathbf{R}^2)$ has no non-trivial projections.

In the authors opinion the corona algebras are destined to be important technical tools in the study of certain "non-local" phenomena in multiplier algebras, thus ultimately to describe certain characteristics for the original algebras. A few supporting cases follow.

Case 5.1. Let T be an element in $\mathbf{B}(\mathfrak{H})$. There it has a spectrum, $\mathrm{sp}(T)$, but often one must also know the *essential spectrum*, $\mathrm{sp}_{ess}(T)$. If for example you wish to estimate the norm distance from T to the groups of invertible or unitary operators on $\mathfrak{H}$, then the lower bound of $\mathrm{sp}_{ess}(|T|)$ enters the formulae. But $\mathrm{sp}_{ess}(T)$ is just the usual spectrum of the image of T in the Calkin algebra.

Case 5.2. The *index* of an operator T in $\mathbf{B}(\mathfrak{H})$ ($= \dim\ker T - \dim\ker T^*$) is an important invariant (for Fredholm operators). But many of its properties are more easily understood by considering the map π from Fredholm operators in $\mathbf{B}(\mathfrak{H})$ onto the group $\mathfrak{G}$ of invertible elements in the Calkin algebra, and the quotient map $j : \mathfrak{G} \to \mathfrak{G}/\mathfrak{G}_0$ with respect to the connected component $\mathfrak{G}_0$ of $\mathfrak{G}$. Clearly $T \to j(\pi(T))$ is a multiplicative map, and once it has been shown that $\mathfrak{G}/\mathfrak{G}_0 = \mathbf{Z}$ it is not difficult to prove that it is, in fact, the index map.

Case 5.3. Let $\mathfrak{A}$ be a unital C^*-algebra and pass to the stable algebra $\mathfrak{A} \otimes \mathfrak{K}$. By a result of J. Mingo (see also the paper by J. Cuntz and N. Higson), the group of invertible elements in $M(\mathfrak{A} \otimes \mathfrak{K})$ is contractible. But the group $\mathfrak{G}$ of invertible elements in the corona C^*-algebra $C(\mathfrak{A} \otimes \mathfrak{K})$ is non-trivial. If $\mathfrak{G}_0$ denotes its connected component, then

$$\mathfrak{G}/\mathfrak{G}_0 = K_1(C(\mathfrak{A} \otimes \mathfrak{K}))$$
$$= K_0(\mathfrak{A} \otimes \mathfrak{K}) = K_0(\mathfrak{A}).$$

Thus for certain "Fredholm" elements in $M(\mathfrak{A} \otimes \mathfrak{K})$ one may define an index with values in $K_0(\mathfrak{A})$. And the beautiful fact is that the two possible ways of doing this give the same result. Thus if T is "Fredholm" in $M(\mathfrak{A} \otimes \mathfrak{K})$ and its kernel and co-kernel are given by projections P and Q in $\mathfrak{A} \otimes \mathfrak{K}$, then

$$[P] - [Q] = j(\pi(T))$$

exactly as in the case 5.2.

§6. Properties of Corona Algebras

A C^*-algebra $\mathfrak{A}$ has the *countable Riesz separation property* $(CRISP)$ if, whenever (A_n) and (B_n) are sequences in $\mathfrak{A}_{sa}$, the first increasing, the other decreasing, such that

$$A_n \leq A_{n+1} \leq B_{n+1} \leq B_n$$

for all n, then there is an element C in $\mathfrak{A}_{sa}$ with $A_n \leq C \leq B_n$ for all n.

Theorem 6.1. *If $\mathfrak{A}$ is a σ-unital C^*-algebra, then its corona algebra $C(\mathfrak{A})$ has the countable Riesz separation property.*

Proof: Consider sequences (A_n) and (B_n) in $C(\mathfrak{A})_{sa}$, as specified above, and let H be a strictly positive element in $\mathfrak{A}$. Let $\pi : M(\mathfrak{A}) \to C(\mathfrak{A})$ denote the quotient map. Assume that for $1 \leq k \leq n$ we have found elements S_k and T_k in $M(\mathfrak{A})_{sa}$ such that

  (i) $S_{k-1} \leq S_k \leq T_k \leq T_{k-1}$,
  (ii) $\pi(S_k) = A_k, \quad \pi(T_k) = B_k$,
  (iii) $\| H(T_k - S_k)H \| \leq k^{-1}$,

for all $k \leq n$. By Lemma 4.2 there is an element T_{n+1} in $M(\mathfrak{A})_{sa}$ with $\pi(T_{n+1}) = B_{n+1}$ and $S_n \leq T_{n+1} \leq T_n$. Using the lemma again, this time on the pair S_n, T_{n+1}, we find an element R in $M(\mathfrak{A})_{sa}$ with $\pi(R) = T_{n+1}$ and $S_n \leq R \leq T_{n+1}$. Now, proceeding exactly as in the proof of Theorem 4.3, we choose an approximative unit (E_λ) for $\mathfrak{A}$ and put

$$S_\lambda = R + (T_{n+1} - R)^{1/2} E_\lambda (T_{n+1} - R)^{1/2}.$$

Note that $\pi(S_\lambda) = A_{n+1}$ for all λ and that

$$S_n \leq R \leq S_\lambda \leq T_{n+1}.$$

Finally

$$\| H(T_{n+1} - S_\lambda)H \| = \left\| H (T_{n+1} - R)^{1/2} (I - E_\lambda) (T_{n+1} - R)^{1/2} H \right\|,$$

and this goes to zero as $\lambda \to \infty$. Thus we can take $S_{n+1} = S_\lambda$ for λ sufficiently large, and have (i), (ii) and (iii) satisfied for $n + 1$, hence by induction for all n.

Working in the universal representation, so that

$$\mathfrak{A} \subset M(\mathfrak{A}) \subset \mathfrak{A}^{**} \subset B(\mathfrak{H}_u),$$

we have $S_n \nearrow S$ and $T_n \searrow T$ for some elements S and T in $\mathfrak{A}^{**}_{sa}$. Moreover, from Theorem 3.2 we see that

$$S \in (\tilde{\mathfrak{A}}_{sa})^m \quad , \quad T \in (\tilde{\mathfrak{A}}_{sa})_m.$$

From (iii) it follows that $H(T - S)H = 0$, and since H is strictly positive this means that $T = S$; whence $T \in M(\mathfrak{A})_{sa}$ by Theorem 3.2. Take $C = \pi(T) \ (= \pi(S))$ and note from (ii) that

$$A_n = \pi(S_n) \leq \pi(S) = C = \pi(T) \leq \pi(T_n) = B_n,$$

for all n, which shows that $C(\mathfrak{A})$ is $CRISP$.

$\square$

In order to prove an asymptotically abelian version of the $CRISP$ theorem, above, we need the following lemma. Actually Arveson's version from [4] would suffice for our purpose; but the commutation problem is intriguing in itself, especially because the best constant (believed to be 1) in the inequality is still unknown.

Lemma 6.2. *If S and T are operators in $B(\mathfrak{H})$ and $T \geq 0$, then for $0 < \beta < 1$ we have*

$$\| [S, T^{\beta}] \| \leq \gamma_m \| [S, T] \|^{\beta} \| S \|^{1-\beta}$$

if $\| [S, T] \| \leq m^{-1}$, where $\gamma_m = m^{\beta} \Pi_{k=2}^{m-1}(1 - k^{-1}\beta)$.

Proof: Without loss of generality we may assume that $\| S \| = 1$ and that $0 \leq T \leq I$. Replacing T by $I - T$ it therefore suffices to prove the inequality

$$(*) \qquad \qquad \|[S, (I - T)^{\beta}]\| \leq \gamma_m \|[S, T]\|^{\beta}$$

assuming that $\| [S, T] \| \leq m^{-1}$. For this we employ the power series expansion

$$(**) \qquad \qquad 1 - (1 - t)^{\beta} = \sum_{k=1}^{\infty} \alpha_n t^n, \quad 0 \leq t \leq 1,$$

where $\alpha_n > 0$ for all n. In fact $\alpha_n = \left| \binom{\beta}{n} \right|$. Define

$$\sigma_n = \sum_{k=1}^{n-1} k \, \alpha_k, \qquad \tau_n = \sum_{k=n}^{\infty} \alpha_k.$$

We claim that

$$(***) \qquad \sigma_n = n(n-1)(1-\beta)^{-1}\alpha_n, \quad \tau_n = n\beta^{-1}\alpha_n.$$

To establish these claims by induction, note first that $\sigma_1 = 0$ and $\tau_1 = 1 - (1-1)^\beta = 1$ as claimed in $(***)$. Assuming now that $(***)$ holds for n, we compute

$$\sigma_{n+1} = \sigma_n + n\alpha_n = (n(n-1)(1-\beta)^{-1} + n)\alpha_n$$
$$= (n+1)n(1-\beta)^{-1}((n+1)^{-1}(n-1) + (n+1)^{-1}(1-\beta))\alpha_n$$
$$= (n+1)n(1-\beta)^{-1}((n+1)^{-1}(n-\beta)\alpha_n)$$
$$= (n+1)n(1-\beta)^{-1}\alpha_{n+1}.$$

Similarly,

$$\tau_{n+1} = \tau_n - \alpha_n = (n\beta^{-1} - 1)\alpha_n = \beta^{-1}(n-\beta)\alpha_n$$
$$= (n+1)\beta^{-1}((n+1)^{-1}(n-\beta)\alpha_n) = (n+1)\beta^{-1}\alpha_{n+1}.$$

Having thus established $(***)$ for all n, we use the power series in $(**)$ to estimate

$$\|[S, (I-T)^\beta]\| = \left\| \sum_{k=1}^{\infty} \alpha_k [S, T^k] \right\|$$

$$\leq \sum_{k=1}^{n-1} \alpha_k \left\| \sum_{j=0}^{k-1} T^j [S, T] T^{k-j-1} \right\| + \sum_{k=n}^{\infty} \alpha_k \| [S, T^k] \|.$$

Since $0 \leq T^k \leq I$ we have

$$\|[S, T^k]\| = \left\| [S, T^k - \tfrac{1}{2}I] \right\|$$
$$\leq 2\|S\| \left\| T^k - \tfrac{1}{2}I \right\| \leq 1.$$

Thus with $\| [S, T] \| = \varepsilon$ we get by $(***)$

$$\|[S, (I-T)^\beta]\| \leq \sum_{k=1}^{n-1} \alpha_k k \varepsilon + \sum_{k=n}^{\infty} \alpha_k$$
$$= \varepsilon \sigma_n + \tau_n = ((1-\beta)^{-1}(n-1)\varepsilon + \beta^{-1})n\alpha_n.$$

Note now that

$$(1-\beta)^{-1}\beta^{-1}n\alpha_n = \frac{(2-\beta)}{2}\frac{(3-\beta)}{3}\cdots\frac{(n-1-\beta)}{n-1} = \prod_{k=2}^{n-1}(1 - k^{-1}\beta).$$

As $1 - x \leq \exp{-x}$ for $0 \leq x \leq 1$ we have

$$\prod_{k=m}^{n-1}(1 - k^{-1}\beta) \leq \exp\left(\sum_{k=m}^{n-1} -k^{-1}\beta\right).$$

As moreover $\sum_{k=m}^{n-1} k^{-1} > \log(nm^{-1})$ we have

$$\prod_{k=m}^{n-1}(1 - k^{-1}\beta) \leq (nm^{-1})^{-\beta} = n^{-\beta}m^{\beta}.$$

Choosing n such that $n - 1 \leq \varepsilon^{-1} < n$ and assuming that $m \leq \varepsilon^{-1}$ we finally have

$$\| [S, (I - T)^{\beta}] \| \leq (1 - \beta)^{-1}\beta^{-1}n\alpha_n$$

$$\leq \prod_{k=2}^{m-1}(1 - k^{-1}\beta)m^{\beta}n^{-\beta} \leq \gamma_m\varepsilon^{\beta},$$

i.e. $(*)$.

$\square$

Corollary 6.3. *If* $\| [S, T] \| = \varepsilon \leq \frac{1}{4}$ *for* $\| S \| = 1$ *and* $0 \leq T \leq I$, *then* $\| [S, T^{1/2}] \| \leq \frac{5}{4}\varepsilon^{1/2}$.

Proof: Take $\beta = \frac{1}{2}$ and $m = 4$ in 6.2.

$\square$

Corollary 6.4. *If* $\| [S, T] \| = \varepsilon \leq \frac{1}{16}$ *for* $\| S \| = 1$ *and* $0 \leq T \leq I$, *then* $\| [S, T^{1/2}] \| \leq 1, 16 \, \varepsilon^{1/2}$.

Proof: Take $\beta = \frac{1}{2}$ and $m = 16$ in 6.2.

$\square$

Remark 6.5. Stirling's formula shows that the lower bound for the constants γ_m is $((1 - \beta)\Gamma(1 - \beta))^{-1}$. For $\beta = \frac{1}{2}$ this number is $2\pi^{-1/2} \sim 1.128$.

Theorem 6.6. *Let* $\mathfrak{A}$ *be a* σ-*unital* C^*-*algebra with corona algebra* $C(\mathfrak{A})$, *and let* (T_n) *be a monotone increasing sequence in* $C(\mathfrak{A})_+$ *and* $\mathfrak{D}$ *a separable subset of* $C(\mathfrak{A})$, *such that* $[D, T_n] \to 0$ *for every* D *in* $\mathfrak{D}$. *Then if* $T_n \leq S$ *for some* S *in* $C(\mathfrak{A}_s)$ *and all* n, *there is a* T *in* $C(\mathfrak{A})_{sa}$, *commuting with* $\mathfrak{D}$, *such that* $T_n \leq T \leq S$ *for all* n.

Proof: Choose (B_n) in $M(\mathfrak{A})$ such that (D_n) is dense in $\mathfrak{D}$, where $D_n = \pi(B_n)$ and π as in 6.1 denotes the quotient map. Passing if necessary to a subsequence we may assume that

(i) $\| [D_k, T_n] \| < 2^{-n}$ for $1 \le k \le n$.

Now use the same arguments as in the proof of Theorem 6.1 to construct an increasing sequence (A_n) in $M(\mathfrak{A})_{sa}$, converging strictly to some A in $M(\mathfrak{A})_{sa}$, such that

(ii) $\pi(A_n) = T_n, \quad T_n \le \pi(A) \le S$

for all n.

Choose an approximate unit (E_n) for $\mathfrak{A}$ which is quasi-central with respect to (A_n), A and (B_k), cf. Theorem 2.2. Thus writing $F_n = (E_n - E_{n-1})^{1/2}$ (and $E_0 = 0$) we may assume, using Lemma 6.2, that

(iii) $\| [A_k, F_n] \| \le 2^{-n}$,

(iv) $\| [B_k, F_n] \| \le 2^{-n}$,

(v) $\| [A, F_n] \| \le 2^{-n}$,

(vi) $\| [F_n[A_n, B_k]F_n] \le 2^{-n}$,

or $1 \le k \le n$. Note that to establish (vi) we use (i), because

$$\| F_n[A_n, B_k]F_n \| \sim \| F_n^2[A_n, B_k] \|$$
$$\le \| (I - E_{k-1})[A_n, B_k] \| \to \| \pi[A_n, B_k] \|$$
$$= \| [T_n, D_k] \| < 2^{-n}.$$

Take $B = \sum F_n A_n F_n$, which belongs to $M(\mathfrak{A})_{sa}$ by Theorem 4.1, and put $T = \pi(B)$. Clearly then $T_n \le T \le S$, because

$$C_n = \sum_{k=1}^{n} F_k A_k F_k + \sum_{k=n+1}^{\infty} F_k A_n F_k \le B,$$

$$B \le \sum_{n=1}^{\infty} F_n A F_n = C,$$

and $A_n - C_n \in \mathfrak{A}$, $A - C \in \mathfrak{A}$ by (iii) and (v), since $\sum F_n^2 = I$. Moreover,

$$[B_k, B] = \sum [B_k, F_n A_n F_n]$$
$$= \sum F_n[B_k, A_n]F_n + G_1 = G_2 + G_1 \in \mathfrak{A},$$

where $G_1 \in \mathfrak{A}$ by (iv) and $G_2 \in \mathfrak{A}$ by (vi). This means that $[D_k, T] = 0$ for all k, i.e. $T \in \mathfrak{D}'$.

$\square$

Combining the last two theorems we see that every corona algebra has the following *asymptotically abelian, countable Riesz separation property* $(AA - CRISP)$:

Corollary 6.7. *If (S_n) and (T_n) are monotone sequences of self-adjoint elements in a corona algebra $C(\mathfrak{A})$, the first increasing, the second decreasing, such that $S_n \leq T_n$ for all n, and such that $[S_n, D] \to 0$ for every D in some separable subset $\mathfrak{D}$ of $C(\mathfrak{A})$, then there is an R in $C(\mathfrak{A})_{sa}$, commuting with $\mathfrak{D}$, such that $S_n \leq R \leq T_n$ for all n.*

§7 *SAW^*-Algebras*

A (unital) C^*-algebra $\mathfrak{A}$ is called an SAW^*-algebra (short for sub-AW^*-algebra) if for any two orthogonal elements S and T in $\mathfrak{A}_+$ there is an element R in $\mathfrak{A}_+$ (which we can then choose with $\| R \| \leq 1$) such that $SR = R$ and $RT = 0$. Applying the condition to the pair R, T we obtain an element Q in $\mathfrak{A}_+$ such that $RQ = 0$ and $QT = 0$. In this more symmetric setting we say that R, Q is a pair of orthogonal *local units* for S and T.

To clarify the relations between the notions of SAW^*-algebras, Rickart algebras and AW^*-algebras we need the concept of (two-sided) *annihilators*. For every subset $\mathfrak{B}$ of a C^*-algebra $\mathfrak{A}$, set

$$\mathfrak{B}^\perp = \{T \in \mathfrak{A} \mid T\mathfrak{B} = \mathfrak{B}T = 0\}.$$

Recall from [20, 1.5.2] that a C^*-subalgebra $\mathfrak{B}$ of $\mathfrak{A}$ is called *hereditary* if $\mathfrak{B} = \mathfrak{L} \cap \mathfrak{L}^*$ for some closed left ideal $\mathfrak{L}$ in $\mathfrak{A}$. Equivalent conditions, see e.g. [20, 3.11.9] are that $\mathfrak{B}_+$ should be an hereditary cone in $\mathfrak{A}_+$, and again that $\mathfrak{B}\mathfrak{A}\mathfrak{B} \subset \mathfrak{B}$, or, finally that $\mathfrak{B} = P\mathfrak{A}^{**}P \cap \mathfrak{A}$ for some projection P in $\mathfrak{A}^{**}$ (which can then be chosen in $(\mathfrak{B}_+)^m$). An hereditary C^*-subalgebra $\mathfrak{B}$ will be σ-unital if and only if it is singly generated, i.e. if $\mathfrak{B} = (H\mathfrak{A}H)^=$ for some positive element H in $\mathfrak{A}$ (which then becomes a strictly positive element in $\mathfrak{B}$). The hereditary algebra $\mathfrak{B}$ is unital if and only if H is a projection (in $\mathfrak{A}$), and $\mathfrak{B}$ is then called a *corner* of $\mathfrak{A}$.

Proposition 7.1. *Suppose that for any two orthogonal, hereditary C^*-subalgebras $\mathfrak{B}$ and $\mathfrak{C}$ of a C^*-algebra $\mathfrak{A}$ there is an element R in $\mathfrak{A}_+$ that is a unit for $\mathfrak{B}$ and annihilates $\mathfrak{C}$. Then $\mathfrak{A}$ is an AW^*-algebra. If the condition only applies when $\mathfrak{B}$ is σ-unital, then $\mathfrak{A}$ is a Rickart algebra. If the condition only applies when both $\mathfrak{B}$ and $\mathfrak{C}$ are σ-unital, then $\mathfrak{A}$ is an SAW^*-algebra.*

Proof: Easy, see [21, Proposition 1].

$\square$

Despite the formal similarity between AW^*-algebras and SAW^*-algebras they are, in reality, quite different. One reason is that the SAW^*-condition, forcing the existence of a local unit R, does not require R to be unique. Applied in Rickart or AW^*-algebras to pairs of the form $\mathfrak{B}$, $\mathfrak{B}^\perp$, the local unit R is unique, and is a projection. By contrast there are commutative examples of SAW^*-algebras with no non-trivial projections (e.g. connected sub-Stonean spaces like $\beta\mathbb{R}^2 \setminus \mathbb{R}$). It should be mentioned, though, that all three classes are somewhat exotic – as C^*-algebras – and will only occur as secondary objects in the theory, viz. as corona algebras. All three classes are plagued with the stability problem: Since the extra structure only concerns commutative subalgebras of $\mathfrak{A}$ it is very hard to decide whether it will persist in matrix algebras over $\mathfrak{A}$, i.e. algebras of the form $\mathsf{M}_n(\mathfrak{A})$. The fact that AW^*-algebras are stable under tensor products with matrix algebras (proved by Berberian in [5], see also [6, §62] may well be the deepest result in the theory. The corresponding problem for SAW^*-algebras is wide open.

Theorem 7.2. *If $\mathfrak{J}$ is a closed, σ-unital ideal in an SAW^*-algebra $\mathfrak{A}$, then*

$$M(\mathfrak{J}) = \mathfrak{A}/\mathfrak{J}^\perp.$$

Proof: Since $\mathfrak{J}$ is isomorphically embedded in $\mathfrak{A}/\mathfrak{J}^\perp$ we may assume that $\mathfrak{J}^\perp = 0$, so that $\mathfrak{J}$ is an essential ideal in $\mathfrak{A}$. By Theorem 3.1 there is then a natural embedding

$$\mathfrak{J} \subset \mathfrak{A} \subset M(\mathfrak{J}).$$

If $\mathfrak{A} \neq M(\mathfrak{J})$ there is a self-adjoint functional of $M(\mathfrak{J})$, of norm one, that annihilates $\mathfrak{A}$. Its Jordan decomposition, cf. [20, 3.2.5] then produces two orthogonal states φ and ψ of $M(\mathfrak{J})$, such that $\varphi - \psi$ annihilates $\mathfrak{A}$. By [20, 3.2.3] we may choose, for any $\varepsilon > 0$, an element E in $M(\mathfrak{J})$ with $0 \leq E \leq I$, such that $\varphi(I - E) < \varepsilon^2$ and $\psi(E) < \varepsilon^2$. Let f and g be continuous functions on $\mathrm{sp}(E)$ such that $0 \leq f, g \leq 1$, and $f(1) = 1$ whereas $f(t) = 0$ for $t < 1 - \varepsilon$, and $g(0) = 1$ whereas $g(t) = 0$ for $t > \varepsilon$. Put $S = f(E)$ and $T = g(E)$ and note that $ST = 0$ (if $\varepsilon < 1/2$). Moreover, $I - S \leq \varepsilon^{-1}(I - E)$ and $I - T \leq \varepsilon^{-1}E$. Consequently $\varphi(I - S) < \varepsilon$ and $\psi(I - T) < \varepsilon$, so that $\varphi(S) > 1 - \varepsilon$ and $\psi(T) > 1 - \varepsilon$.

Let H be a strictly positive element for $\mathfrak{J}$. Since SH^2S and TH^2T are orthogonal elements in the SAW^*-algebra $\mathfrak{A}$, there are orthogonal local units R, Q in $\mathfrak{A}_+$ for SH^2S and TH^2T. Thus $(I - R)SH^2S(I - R) = 0$, whence $(I - R)SH = 0$. Since H is strictly positive in $\mathfrak{J}$ it follows that $(I - R)S = 0$. In particular $R \geq S$. Similarly we show that $Q \geq T$. But now we conclude that since $\varphi \perp \psi$,

$$1 \geq (\varphi + \psi)(R + Q) \geq \varphi(R) + \psi(Q)$$
$$\geq \varphi(S) + \psi(T) \geq 2(1 - \varepsilon).$$

Having chosen $\varepsilon < 1/2$ we reach a contradiction.

$\square$

Remark 7.3. In the commutative case, $\mathfrak{A} = C(X)$, we see that $\mathfrak{A}$ is an SAW^*-algebra if and only if X is a sub-Stonean space (also known as an F-space), which by definition means that no two disjoint, σ-compact subsets of X can have any common boundary points. Using the fact that the multiplier algebra is the algebraic counterpart of the Stone-Čech compactification, it follows that the contents of Theorem 7.2 is the formula

$$\beta(Y) = Y^- \subset X$$

for every open, σ- compact subset Y of the sub-Stonean space X.

Theorem 7.4. *Every C^*-algebra $\mathfrak{A}$ that satisfies the countable Riesz separation property is an SAW^*-algebra.*

Proof: Let S and T be orthogonal, positive elements in $\mathfrak{A}$, say with norms less than or equal to one. Then

$$S^{1/n} \le (I - T)^n$$

for all n by spectral theory. Since $(S^{1/n})$ is increasing and $((I - T)^n)$ decreasing it follows from $CRISP$ that for some element R in $\mathfrak{A}$ we have $S^{1/n} \le R \le (I - T)^n$ for all n. In particular, $0 \le R \le I$. We have

$$\| S(I - R) \| \le \| S(I - S^{1/n}) \|,$$
$$\| TR \| \le \| T(I - T)^n \|,$$

for all n, whence $SR = S$ and $RT = 0$, as desired.

$\square$

Corollary 7.5. *Every corona algebra is an SAW^*-algebra.*

Remark 7.6. By a delightful imitation of Urysohn's lemma it is proved in [27] that a compact Hausdorff space X is sub-Stonean (i.e. $C(X)$ is an SAW^*-algebra) if and only if $C(X)$ has $CRISP$. This, however is unlikely to hold in the non-commutative case, where $CRISP$ seems far stronger than SAW^*.

Theorem 7.7. *Let $\mathfrak{A}$ be a σ-unital C^*-algebra with corona algebra $C(\mathfrak{A})$. Then for each hereditary, σ-unital C^*-subalgebra $\mathfrak{B}$ of $C(\mathfrak{A})$ we have*

$$\mathfrak{B}^{\perp\perp} = \mathfrak{B}.$$

Proof: Since evidently $\mathfrak{B} \subset \mathfrak{B}^{\perp\perp}$, it suffices to show that if $S \notin \mathfrak{B}$, $0 \leq S \leq I$, there is a T in $\mathfrak{B}^{\perp}$ such that $ST \neq 0$. Towards this end, let $\pi : M(\mathfrak{A}) \to C(\mathfrak{A})$ be the quotient map, and choose H in $M(\mathfrak{A})_+$ such that $\pi(H)$ is strictly positive in $\mathfrak{B}$. Adding, if necessary, a strictly positive element from $\mathfrak{A}$ to H we may assume that the (σ-unital) hereditary C^*-subalgebra $\mathfrak{D}$ of $M(\mathfrak{A})$, generated by H, contains $\mathfrak{A}$. Note that $\pi(\mathfrak{D}) = \mathfrak{B}$ by [20, 1.5.11]. Since $S \notin \mathfrak{B}$ we can find A in $M(\mathfrak{A})$, $0 \leq A \leq I$ with $\pi(A) = S$ such that $A \notin \mathfrak{D}$. If $\mathfrak{L}$ denotes the closed left ideal of $M(\mathfrak{A})$ for which $\mathfrak{D} = \mathfrak{L} \cap \mathfrak{L}^*$, we therefore have $A \notin \mathfrak{L}$ so that

$$(*) \qquad\qquad \mathrm{dist}(A, \mathfrak{L}) > \alpha$$

for some $\alpha > 0$.

Let (E_n) and (D_n) be approximate units for $\mathfrak{A}$ and $\mathfrak{D}$, respectively. By Theorem 3.2 we may assume that $E_n E_{n-1} = E_n$ for all n, and that

$$(**) \qquad\qquad \| [A, E_n] \| \leq 2^{-n}, \quad \| [H, E_n] \| \leq 2^{-n}$$

for all n. Suppose now that for some n we had

$$\|(I - D_n)(E_m - E_{n-1})^{1/2} A\| \leq \alpha$$

for all $m \geq n$. Since (E_m) converges strongly to I in $\mathfrak{A}^{**}$ this would mean that

$$\| (I - D_n)(I - E_{n-1})^{1/2} A \| \leq \alpha.$$

However, since $\mathfrak{A} \subset \mathfrak{D}$,

$$A - (I - D_n)(I - E_{n-1})^{1/2} A \in \mathfrak{D} A \subset \mathfrak{L}^*,$$

contradicting $(*)$. Passing if necessary to a subsequence of (E_n) (a process that doesn't disturb $(**)$) we may assume that

$$(***) \qquad\qquad \| (I - D_n)(E_n - E_{n-1})^{1/2} A \| > \alpha$$

for all n.

Put $F_n = (E_n - E_{n-1})^{1/2}$, and define

$$C = \sum F_n (I - D_n)^2 F_n.$$

By Theorem 4.1 we know that $C \in M(\mathfrak{A})$. To show that $CH \in \mathfrak{A}$, consider the element

$$B = \sum F_n(I - D_n)^2 H \, F_n.$$

Since $\| (I - D_n)H \| \to 0$ it follows that $B \in \mathfrak{A}$, since $(F_n) \subset \mathfrak{A}$. However, using Corollary 6.3 in conjunction with $(**)$ we see that $CH - B \in \mathfrak{A}$ and thus $CH \in \mathfrak{A}$. Since $\mathfrak{D} = (HM(\mathfrak{A})H)^=$ this implies that $C\mathfrak{D} \subset \mathfrak{A}$. Thus with $T = \pi(C)$ we know that $T \in \mathfrak{B}^{\perp}$. To show that $ST \neq 0$ is equivalent to proving that $CA \notin \mathfrak{A}$. But if it were, then for a suitable large m we would have

$$\| (I - E_m)ACA(I - E_m) \| < \alpha^2$$

and thus also

$$\| A(I - E_m)C(I - E_m)A \| < \alpha^2$$

by $(**)$. However, since $F_n(I - E_m) = F_n$ for $n \geq m + 2$, this means that

$$\alpha^2 > \| A(I - E_m)F_n(I - D_n)^2 F_n(I - E_m)A \|$$
$$= \| (I - D_n)F_n A \|^2,$$

which contradicts $(***)$. Thus $CA \notin \mathfrak{A}$, as desired.

$\square$

Remark 7.8. In the commutative case, where $\mathfrak{A} = C_0(X)$ for some locally compact, σ-compact Hausdorff space, which implies that $C(\mathfrak{A}) = C(\beta(X) \setminus X)$, the content of the theorem above is that every open, σ-compact subset Y of the corona space $\beta(X) \setminus X$ is regularly embedded, i.e. is equal to the interior of its closure, see [12].

Even in the classical case, where $\mathfrak{A} = \mathfrak{K}$, the algebra of compact operators on a separable Hilbert space $\mathfrak{H}$, the result appears to be new. The following is an easy reformulation of it.

Corollary 7.9. *If S and T are positive operators in $\mathsf{B}(\mathfrak{H})$ and T is non-singular, then either $S \leq f(T)$ for some continous function f on $\mathrm{sp}(T)$ with $f(0) = 0$ (i.e. $S \in (T\mathsf{B}(\mathfrak{H})T)^=$), or else there is an operator $R \geq 0$ such that $TR \in \mathfrak{K}$, but $SR \notin \mathfrak{K}$.*

We finally mention a regularity property for corona algebras, related to annihilators. According to Tomita (and popularized by Effros [11]) a projection P in the second dual $\mathfrak{A}^{**}$ of a unital C^*-algebra $\mathfrak{A}$ is *regular* if

$$\| AP \| = \inf\{\| A - B \| \mid B \in \mathfrak{A}, \ BP = 0\}$$

for every A in $\mathfrak{A}$. With $\bar{P}$ as the smallest projection in $(\mathfrak{A}_+)_m$, majorizing P – a fascinating attempt at a non-commutative version of the closure operation – Akemann shows in [1] that regularity means that $\| AP \| = \| A\bar{P} \|$ for every A in $\mathfrak{A}$. In [21, 19] the author showed that regularity also is equivalent to the condition: if $A \in \mathfrak{A}_+$ and $A \geq P$ then $A \geq \bar{P}$. The point being that in the absense of regularity the majorants for P in $\mathfrak{A}_+$ will not have a minimum in $\mathfrak{A}^{**}$.

If P is the support projection (in $(\mathfrak{A}_+)^m$) for an hereditary C^*-subalgebra $\mathfrak{B}$ of $\mathfrak{A}$, the regularity of P can be phrased in terms of $\mathfrak{B}$ and $\mathfrak{B}^\perp$. Thus [18, 21] takes the following form.

Proposition 7.10. *Let $\mathfrak{B}$ be an hereditary, σ-unital C^*-subalgebra of a corona algebra $C(\mathfrak{A})$. Then*

$$\sup\{\| AB \| \mid B \in \mathfrak{B},\ 0 \leq B \leq I\}$$
$$= \inf\{\| A - C \| \mid C \in \mathfrak{B}^\perp\}$$

for every A in $C(\mathfrak{A})_{sa}$.

§8. Kasparov's Technical Theorems

A celebrated result from C^*-algebraic K-Theory, known as *Kasparov's technical theorem* [14, §3 Theorem 3] can be derived from the $AA - CRISP$ results in §6. The trick is to visualize KTT as an asymptotically abelian version of the SAW^*-condition. It was performed by N. Higson, but many specialists, including Cuntz and Skandalis, were aware of the possibility.

To ease the formulation we say that a subset $\mathfrak{D}$ *derives* a (hereditary) C^*-subalgebra $\mathfrak{B}$, if $[D, B] \in \mathfrak{B}$ for every B in $\mathfrak{B}$ and D in $\mathfrak{D}$.

Theorem 8.1. *(KTT) Let $\mathfrak{B}$ and $\mathfrak{C}$ be orthogonal, σ-unital hereditary C^*-subalgebras of a corona algebra $C(\mathfrak{A})$. Assume further that $\mathfrak{D}$ is a separable subset of $C(\mathfrak{A})$ deriving $\mathfrak{B}$. There is then an element R in $C(\mathfrak{A})$, $0 \leq R \leq I$, commuting with $\mathfrak{D}$, which is a unit for $\mathfrak{B}$ and annihilates $\mathfrak{C}$.*

Proof: Let S and T be strictly positive elements for $\mathfrak{B}$ and $\mathfrak{C}$, respectively. Thus, as mentioned above, $\mathfrak{B} = (SC(\mathfrak{A})S)^=$. It follows that $(S^{1/n})$ is an approximate unit for $\mathfrak{B}$. For each B in $\mathfrak{B}$ and D in $\mathfrak{D}$ we therefore have

$$\lim[S^{1/n}, D]B = \lim(S^{1/n}[D, B] + [S^{1/2}B, D])$$
$$= [D, B] + [B, D] = 0.$$

This means that the sequence $([S^{1/n}, D])$ converges strictly to zero in $\mathfrak{B}$, hence also weakly to zero. As in the proof of Theorem 2.1 we conclude from the Hahn-Banach theorem that $\| [E_n, D] \| \to 0$, where each E_n is a convex combination of elements from $(S^{1/n})$. Working by induction, using a dense sequence (D_k) in $\mathfrak{D}$, we can therefore construct an approximate unit (E_n) for $\mathfrak{B}$ such that $E_n \geq S^{1/n}$ and $\| [E_n, D_k] \| < n^{-1}$ for $1 \leq k \leq n$.

Put $F_n = (I - T)^n$ and observe that (F_n) is a decreasing sequence with $E_n \leq F_n$ for all n, because $ST = 0$. By Corollary 6.7 there is an element R in $C(\mathfrak{A})$, $0 \leq R \leq I$, commuting with $\mathfrak{D}$, such that $E_n \leq R \leq F_n$ for all n. In our case this means that

$$S^{1/n} \leq R \leq (I - T)^n$$

for all n; and it follows from spectral theory that R is a unit for S and annihilates T. Since S and T are strictly positive elements in $\mathfrak{B}$ and $\mathfrak{C}$, respectively, this means that $\mathfrak{B}(I - R) = 0$ and $\mathfrak{C}R = 0$, as desired.

$\square$

Remark 8.2. Kasparov formulates his results upstairs in the multiplier algebra $M(\mathfrak{A})$ and does not use the concepts of hereditary algebras and strictly positive elements (which, to some extent, makes the statements very technical). The quotient formulation in 8.1 can be found in Blackadar's book [7, p. 123] and is originally due to N. Higson.

The following equivalent form of KTT treats the subalgebras symmetrically, and is phrased entirely in terms of orthogonality relations. It might therefore conceivably be valid in any SAW^*-algebra.

Theorem 8.3. *Let $\mathfrak{D}$ be a separable unital C^*-subalgebra of a corona algebra $C(\mathfrak{A})$, and assume that $S\mathfrak{D}T = 0$ for some elements S and T in $C(\mathfrak{A})$. There is then an element R in $C(\mathfrak{A})$, $0 \leq R \leq I$, commuting with $\mathfrak{D}$, such that $SR = S$ and $RT = 0$.*

Proof: Let (D_k) be a dense sequence in the unit ball of $\mathfrak{D}$, say with $D_1 = I$, and put $B = \sum 2^{-n} D_n^* S^* S D_n$. Since $\mathfrak{D}$ is an algebra we have $B\mathfrak{D}T = 0$. Now, if $D \in \mathfrak{D}_{sa}$, then

$$2i[D, B] = (I + iD)B(I - iB) - (I - iD)B(I + iD)$$
$$\leq (I + iD)B(I - iD) + (I - iD)B(I + iD) = 2(B + DBD),$$

and similarly $2i[D, B] \geq -2(B + DBD)$. This means that $[D, B]$ belongs to the hereditary C^*-subalgebra $\mathfrak{B}$ of $C(\mathfrak{A})$ generated by B, because $DBD \in \mathfrak{B}$ by our construction of B. Thus $\mathfrak{D}$ derives $\mathfrak{B}$. Taking $\mathfrak{C} = (TC(\mathfrak{A})T^*)^=$ we are back at the assumption in Theorem 8.1, and the desired conclusion follows.

$\square$

§9. Weak Polar Decomposition

We say that a subset $\mathfrak{S}$ of a (unital) C^*-algebra $\mathfrak{A}$ has *weak polar decomposition* if each element T in $\mathfrak{S}$ can be written in the form $T = V|T|$ for some element V in $\mathfrak{S}$ with $\| V \| \leq 1$. Here, as usual, $|T| = (T^*T)^{1/2}$. If V can always be chosen unitary, we say that $\mathfrak{S}$ has *unitary polar decomposition*.

Proposition 9.1. *A C^*-algebra $\mathfrak{A}$ is an SAW^*-algebra if and only if $\mathfrak{A}_{sa}$ has weak polar decomposition. If, moreover, $\mathsf{M}_2(\mathfrak{A})$ is SAW^*, then $\mathfrak{A}$ has weak polar decomposition; and if $\mathsf{M}_4(\mathfrak{A})$ is SAW^*, there is for each pair A, B in $\mathfrak{A}$, such that $A^*A \leq B^*B$, an element W in $\mathfrak{A}$, with $\| W \| \leq 1$, such that $A = WB$.*

Proof: If $\mathfrak{A}$ is SAW^* and $T \in \mathfrak{A}_{sa}$, consider the orthogonal decomposition $T = T_+ - T_-$. There is then an element R in $\mathfrak{A}$, $0 \leq R \leq I$ such that $T_+ R = T_+$ and $RT_- = 0$. Put $V = 2R - I$ and note that $V = V^*$ and $\| V \| \leq 1$. Moreover,

$$V|T| = (2R - I)(T_+ + T_-) = T_+ - T_- = T.$$

Conversely, if $\mathfrak{A}_{sa}$ has weak polar decomposition, consider an orthogonal pair S, T in $\mathfrak{A}_+$. By assumption

$$S - T = V|S - T| = V(S + T)$$

for some V in $\mathfrak{A}_{sa}$ with $\| V \| \leq 1$. Put $R = \frac{1}{2}(I + V)$, so that $0 \leq R \leq I$ and $I - R = \frac{1}{2}(I - V)$. It is immediate to verify that $SR = S$ and $RT = 0$.

If $\mathsf{M}_2(\mathfrak{A})$ is SAW^* and $T \in \mathfrak{A}$, we apply the preceding result to the self-adjoint element $e_{21} \otimes T + e_{12} \otimes T^*$ in $\mathsf{M}_2(\mathfrak{A})$, to obtain a self-adjoint matrix satisfying the decomposition equation

$$\begin{pmatrix} 0 & T^* \\ T & 0 \end{pmatrix} = \begin{pmatrix} A & V^* \\ V & B \end{pmatrix} \left| \begin{pmatrix} 0 & T^* \\ T & 0 \end{pmatrix} \right| = \begin{pmatrix} A & V^* \\ V & B \end{pmatrix} \begin{pmatrix} |T| & 0 \\ 0 & |T^*| \end{pmatrix}.$$

Direct computation shows that $T = V|T|$ and $\| V \| \leq 1$.

Finally, if $\mathsf{M}_4(\mathfrak{A})$ is SAW^*, and $A^*A \leq B^*B$, consider the elements

$$T = \begin{pmatrix} (B^*B - A^*A)^{1/2} & 0 \\ A & 0 \end{pmatrix}, S = \begin{pmatrix} |B| & 0 \\ 0 & 0 \end{pmatrix}$$

73

in $\mathsf{M}_2(\mathfrak{A})$, and note that $T^*T = S^2$, i.e. $|T| = S$. Since $\mathsf{M}_4(\mathfrak{A}) = \mathsf{M}_2(\mathsf{M}_2(\mathfrak{A}))$ we can apply the preceding result to obtain a matrix $C = (C_{ij})$ in $\mathsf{M}_2(\mathfrak{A})$ with $\| C \| \leq 1$ such that $T = CS$. Since moreover $B = V|B|$ for some V in $\mathfrak{A}$ with $\| V \| \leq 1$ it follows that with $W = C_{21}V^*$ we have

$$WB = C_{21}V^*V|B| = C_{21}|B| = A.$$

$\square$

Since $M(\mathsf{M}_n(\mathfrak{A})) = \mathsf{M}_n(M(\mathfrak{A}))$ for any σ-unital C^*-algebra $\mathfrak{A}$ and every n, it follows that $\mathsf{M}_n(C(\mathfrak{A})) = C(\mathsf{M}_n(\mathfrak{A}))$, so that the property of being a corona algebra is stable. Thus all the $AA - CRISP$ and SAW^*-properties proved for $C(\mathfrak{A})$ in §6–§8 are valid for matrix algebras over $C(\mathfrak{A})$. It follows immediately from Proposition 9.1 that every corona algebra has weak polar decomposition. However, appealing to the asymptotically abelian techniques from Theorem 8.3 we obtain a considerably stronger version of 9.1.

Theorem 9.2. *If $\mathfrak{D}$ is a separable C^*-subalgebra of a corona algebra $C(\mathfrak{A})$, and $T \in C(\mathfrak{A})$ such that $T \in \mathfrak{D}'$, then $T = V|T|$ for some V in $C(\mathfrak{A}) \cap \mathfrak{D}'$ with $\| V \| \leq 1$. If T is self-adjoint, V can be chosen self-adjoint, and if T is normal, V can be chosen normal and commuting with T and T^*.*

Proof: We may evidently assume that $I \in \mathfrak{D}$. If $T = T^*$, then $T_+\mathfrak{D}T_- = 0$, and thus by Theorem 8.3 we have $T_+R = T_+$ and $RT_- = 0$ for some R in $C(\mathfrak{A}) \cap \mathfrak{D}'$ with $0 \leq R \leq I$. As in 9.1 we now let $V = 2R - I$ to obtain the desired decomposition $T = V|T|$. In the general case we pass to $\mathsf{M}_2(C(\mathfrak{A}))$ exactly as in the proof of 9.1., and apply the self-adjoint result with respect to the diagonal algebra $I \otimes \mathfrak{D}$. The resulting self-adjoint matrix will then have entries from $\mathfrak{D}'$, as desired.

If T is normal and $T \in \mathfrak{D}'$, we may assume from the outset that $T \in \mathfrak{D}$. By the previous result this gives a decomposition $T = V|T|$ with V in $\mathfrak{D}'$. In particular, V commutes with T. Consequently

$$V^*V|T| = V^*T = TV^* = V|T|V^* = VV^*|T|,$$

so that $(V^*V - VV^*)|T| = 0$. Applying Theorem 8.3 to this orthogonality relation, and with $\mathfrak{D}$ enlarged with V, we obtain an element R in $\mathfrak{D}'$, $0 \leq R \leq I$, which is a unit for $|T|$ and annihilates $V^*V - VV^*$. Since R commutes with V, this means that the element $U = VR$ is normal. Plainly $U \in \mathfrak{D}'$ and

74

$$U|T| = VR|T| = V|T| = T.$$

$\square$

It is not, in general, possible to write $T = U|T|$ with U unitary, even if T is normal. Consider for example a locally compact, σ-compact Hausdorff space X, such that $\dim(X \setminus Z) \geq 2$ for every compact subset Z of X. Then the corona set $Y = \beta X \setminus X$ has $\dim Y \geq 2$ by [12, 3.6]. By definition of the Čech covering dimension there is therefore a unitary function $f_0 : Z \to S^1$, defined on a closed subset Z of Y, that has no continuous extension as a unitary function on all of Y. However, by Tietze's extension theorem there is an extension f in $C(Y)$ with $\| f \| = 1$ (identifying S^1 with the unit circle in $\mathbb{C}$). In the corona algebra $C(Y)$ the normal element f has no unitary polar decomposition $f = u|f|$, because then $u|Z = f|Z = f_0$, so that u would be a unitary extension of f_0.

If T is a self-adjoint element in a corona algebra $C(\mathfrak{A})$ then $T = V|T|$ for some V in $C(\mathfrak{A})_{sa}$ with $\| V \| = 1$. Since $V^*V = V^2$ is a unit for $|T|$ we have $T = U|T|$, where $U = V + i(1 - V^2)^{1/2}$ is unitary in $C(\mathfrak{A})$. Even so, we can in general not expect to find a self-adjoint unitary (i.e. a symmetry) U such that $T = U|T|$. The reason is that then $\frac{1}{2}(I + U)$ is a projection; and there are corona algebras with no non-trivial projections. One such is $C(\mathfrak{A}) = C(\beta \mathbb{R}^2 \setminus \mathbb{R}^2)$, because the corona space $\beta \mathbb{R}^2 \setminus \mathbb{R}^2$ is connected by [12, 3.5]. The same holds for the one-dimensional space $\beta \mathbb{R}_+ \setminus \mathbb{R}_+$.

These obstructions suggest that we consider corona algebras with low "dimension". The proper dimension concept in this connection is the Bass stable rank, see [13], [24], [25]; and the lowest rank –one – simply means that the group $GL(C(\mathfrak{A}))$ of invertible elements in $C(\mathfrak{A})$ is dense.

Lemma 9.3. *If T is an element in a unital C^*-algebra $\mathfrak{A}$, admitting a weak polar decomposition $T = V|T|$, such that $\mathrm{dist}(V, GL(\mathfrak{A})) < 1$, then T has a unitary polar decomposition.*

Proof: If $\mathrm{dist}(GL(\mathfrak{A}), V) < \alpha < 1$ and $f \in C(\mathbb{R})$ such that $f(t) = 0$ for $t \leq \alpha$, but $f(1) = 1$, then by [22, Corollary 8] there is a unitary U in $\mathfrak{A}$ such that

$$V f(|V|) = U|V|f(|V|).$$

Moreover, since $V^*V|T| = |T|$, $f(|V|)|T| = |T|$. Consequently,

$$T = V|T| = Vf(|V|)|T| = U|V|f(|V|)|T| = U|T|.$$

$\square$

Proposition 9.4. *If a unital C^*-algebra $\mathfrak{A}$ has unitary polar decomposition, then $\mathfrak{A}$ is an SAW^*-algebra with $GL(\mathfrak{A})$ dense in $\mathfrak{A}$. Conversely, if $GL(\mathfrak{A})$ is dense in $\mathfrak{A}$ and $\mathsf{M}_2(\mathfrak{A})$ is an SAW^*-algebra, then $\mathfrak{A}$ has unitary polar decomposition.*

Proof: Since unitary polar decomposition implies weak polar decomposition we see from 9.1 that $\mathfrak{A}$ is an SAW^*-algebra. Moreover, since $\mathfrak{A}_+ \subset (GL(\mathfrak{A}))^=$, every product $T = U|T|$, with U unitary, belongs to the closure of $GL(\mathfrak{A})$, which therefore equals $\mathfrak{A}$.

Conversely, if $\mathsf{M}_2(\mathfrak{A})$ is SAW^*, then $\mathfrak{A}$ admits weak polar decomposition by Proposition 9.1 and if furthermore $GL(\mathfrak{A})$ is dense in $\mathfrak{A}$ it follows from Lemma 9.3 that $\mathfrak{A}$ admits unitary polar decomposition.

$\square$

Corollary 9.5. *A corona algebra $C(\mathfrak{A})$ admits unitary polar decomposition if and only if $GL(C(\mathfrak{A}))$ is dense in $C(\mathfrak{A})$.*

Remark 9.6. The result above would be more applicable if we knew how to determine the density of invertibles in $C(\mathfrak{A})$ from properties of the original algebra $\mathfrak{A}$. The commutative test case works fine: If $\mathfrak{A} = C_0(X)$ for some locally compact, σ-compact Hausdorff space X then by [5] $\dim(\beta X \setminus X) = \inf \dim(X \setminus Y)$ the infimum being taken over all open, relatively compact subsets Y of X. Thus $GL(C(\beta X \setminus X))$ is dense if and only if $\dim \beta X \setminus X \leq 1$, and this will happen if $\dim X \leq 1$. By contrast, if $\mathfrak{A} = \mathfrak{K}$, then certainly we are in a zero-dimensional situation, no matter what concept of rank or dimension is being used. But the invertible elements in the Calkin algebra are not dense – consider the image of an isometry in $\mathsf{B}(\ell^2)$ with infinite-dimensional co-rank.

One result that holds in the Calkin algebra (but which is largely undecided in the commutative case) is that every element in the closure of the invertible elements has a unitary polar decomposition. This could very well be true in all corona algebras. The next results can be regarded as weak versions of the general statement.

Proposition 9.7. *If T is an element in a corona algebra $C(\mathfrak{A})$, such that $T + T^* \geq 0$, then $T = U|T|$ for some unitary U in $C(\mathfrak{A})$ with $U + U^* \geq 0$.*

Proof: We first show that T has a weak polar decomposition $T = V|T|$ with $V + V^* \geq 0$. Choose a preimage S for T in $M(\mathfrak{A})$ with $S + S^* \geq 0$. Then $S + n^{-1}I$ is invertible in $M(\mathfrak{A})$ for every n, so that $S + n^{-1}I = U_n|S + n^{-1}I|$ for some unitary U_n

$(= (S + n^{-1}I)|S + n^{-1}I|^{-1})$. Put $H_n = |S + n^{-1}I|$ and note that

$$U_n H_n + H_n U_n^* = S + S^* + 2n^{-1}I \geq 2n^{-1}I.$$

This inequality is stable under application of the inner automorphisms $\mathrm{Ad}\, U_n^k$ for all k in $\mathbb{Z}$, so replacing H_n with $K_m = (2m + 1)^{-1} \sum_{k=-m}^{m} U_n^k H_n U_n^{-k}$ we still have

$$U_n K_m + K_m U_n^* \geq 2n^{-1}I,$$

and $0 \leq K_m \leq H_n$. Let K be a weak limit point for the sequence (K_m). Then K commutes with U_n so that we obtain the estimate

$$\|H_n\|(U_n + U_n^*) \geq U_n K + K U_n^* \geq 2n^{-1}I.$$

In particular, $U_n + U_n^* \geq 0$, cf. [16, 6.1] or [30]. Now let (E_n) be an approximate unit for $\mathfrak{A}$ which is quasi-central with respect to (U_n) and S, and define

$$W = \sum (E_n - E_{n-1})^{1/2} U_n (E_n - E_{n-1})^{1/2}.$$

Then $W \in M(\mathfrak{A})$ by Theorem 4.1, with $\|W\| \leq 1$, and clearly $W + W^* \geq 0$. With $F_n = (E_n - E_{n-1})^{1/2}$ we get

$$W|S| - S = \sum F_n U_n F_n |S| - F_n^2 S$$
$$\sim \sum F_n (U_n |S| - S) F_n \sim \sum F_n (U_n |S + n^{-1}I| - (S + n^{-1}I)) F_n = 0,$$

where $\sim$ denotes equality modulo elements from $\mathfrak{A}$. Consequently, with V as the image of W in $C(\mathfrak{A})$ we have $T = V|T|$ with $\|V\| \leq 1$ and $V + V^* \geq 0$, as desired.

Write $T = H + iK$ and $V = A + iB$, with H, K, A, and B in $(\mathfrak{A})_{sa}$. Note that each element R such that $TR = 0$ will satisfy

$$0 = R^* T R = R^* H R + i R^* K R,$$

whence $R^* H R = R^* K R = 0$. However, since $H \geq 0$ this means that $HR = 0$, whence $iKR = (T - H)R = 0$. We conclude that every one-sided annihilator of T will annihilate both H and K, and therefore also T^* $(= H - iK)$, and thus be a two-sided annihilator. It follows that both $I - V^* V$ and $I - V V^*$ annihilates $|T|$, which means that $\frac{1}{2}(V^* V + V V^*)$ is a unit for $|T|$ and $V^* V - V V^*$ annihilates $|T|$. Since $V^* V + V V^* = 2(A^2 + B^2)$, $V^* V - V V^* = 2i(AB - BA)$ and $A \geq 0$, we conclude from the equations $A^2 |T| = (I - B^2)|T|$ and $AB|T| = BA|T|$ that $A|T| = (I - B^2)^{1/2}|T|$. Define $U = (I - B^2)^{1/2} + iB$. Then U is unitary with $U + U^* \geq 0$. Moreover

$$T = V|T| = (A + iB)|T| = (I - B^2)^{1/2}|T| + iB|T| = U|T|,$$

as claimed.

$\square$

Corollary 9.8. *Every element T in a corona algebra that has the form $T = \lambda(U + V)$ for some λ in $\mathbb{C}$ and unitaries U and V, admits a unitary polar decomposition.*

Proof: If $T = \lambda(U + V)$ then $\lambda^{-1}U^*T = I + U^*V$, and this element clearly has positive real part. Thus

$$I + U^*V = W|I + U^*V| = W|U + V|$$

for some unitary W by Proposition 9.7., i.e.

$$T = \lambda U(I + U^*V) = \lambda UW|U + V| = (\lambda|\lambda|^{-1}UW)|T|.$$

$\square$

§10. Derivations and Morphisms

We know from Sakai's classical results that every (bounded) derivation of a von Neumann algebra or of a simple C^*-algebra is inner, see e.g. [20, 8.6.6 & 8.6.10]. The same is probably true in corona algebras. At the present time, however, we can only prove a local version of this result.

Theorem 10.1. *Let δ be a bounded derivation of a separable C^*-subalgebra $\mathfrak{B}$ of some corona algebra $C(\mathfrak{A})$. There is then an element S in $C(\mathfrak{A})$ such that $\delta(B) = [S, B]$ for every B in $\mathfrak{B}$. If $\delta^* = \delta$ we can choose S such that $0 \leq S \leq \|\delta\|$.*

Proof: Since $\delta = \delta_1 + i\delta_2$, where $\delta_j^* = -\delta_j$ for $j = 1, 2$, we may as well assume that $\delta^* = -\delta$. Now choose a separable C^*-subalgebra $\mathfrak{D}$ of $M(\mathfrak{A})$ such that $\pi(\mathfrak{D}) = \mathfrak{B}$, where π denotes the quotient map. Let (A_k) be a dense sequence in $\mathfrak{D}$ and let (B_k) be its (dense) image in $\mathfrak{B}$ under π. Recall from [20, 8.6.12] that there is an increasing sequence (S_n) in $\mathfrak{B}_+$, bounded by $\|\delta\|$, such that

(i)
$$\|\delta(B_k) - [S_n, B_k]\| < 2^{-n}$$

for all n and k with $k \leq n$. We claim that there is a corresponding increasing sequence (T_n) in $\mathfrak{D}_+$, with $0 \leq T_n \leq \|\delta\|$, $\pi(T_n) = S_n$, such that

(ii)
$$\|[T_n - T_{n-1}, A_k]\| < 2^{-n+2}$$

for $k \leq n$. Suppose that we have already chosen $T_1, \dots, T_{n-1}$ subject to these conditions. By Lemma 4.2 there is then an element T in $M(\mathfrak{A})$, with $\pi(T) = S_n$, such that $T_{n-1} \leq$

$T \le \|\delta\| I$. In order also to satisfy (ii), take an approximate identity (E_λ) for $\mathfrak{A}$ which is quasi-central with respect to (A_k) and $(T - T_{n-1})^{1/2}$, cf. Theorem 2.1. Consider the element

$$T_\lambda = T - (T - T_{n-1})^{1/2} E_\lambda (T - T_{n-1})^{1/2},$$

and note that $T_{n-1} \le T_\lambda \le T$ and $\pi(T_\lambda) = S_n$ for all λ. Moreover, for $1 \le k \le n$ we have

$$\limsup \|[T_\lambda - T_{n-1}, A_k]\|$$
$$= \limsup \|[(T - T_{n-1})^{1/2}(I - E_\lambda)(T - T_{n-1})^{1/2}, A_k]\|$$
$$= \limsup \|[T - T_{n-1}, A_k](I - E_\lambda)\|$$
$$= \|\pi[T - T_{n-1}, A_k]\| < 2^{-n} + 2^{-n+1} < 2^{-n+2}$$

by (i) and [20, 1.5.4]. Taking therefore $T_n = T_\lambda$ for λ sufficiently large we have $\|[T_n - T_{n-1}, A_k]\| < 2^{-n+2}$ for $1 \le k \le n$. By induction the sequence (T_n) can thus be constructed, as claimed.

Define $\delta_0(A_k) = \lim[T_n, A_k]$, cf. (ii). Then δ_0 extends uniquely to a derivation of $\mathfrak{D}$ with $\delta_0^* = -\delta_0$ and $\|\delta_0\| \le \|\delta\|$. Moreover,

$$\pi(\delta_0(A_k)) = \lim \pi[T_n, A_k] = \lim[S_n, B_k] = \delta(B_k),$$

which shows that the image of δ_0 is δ, or, equivalently, that δ_0 is a lifting of δ, cf. [20, 8.6.15].

Choose a countable approximate unit (E_n) for $\mathfrak{A}$ which is quasi-central for (A_k) and for $(\delta_0(A_k))$. Specifically we may assume, using Lemma 6.2, that with $F_n = (E_n - E_{n-1})^{1/2}$ (and $E_0 = 0$) we have

(iii) $$\|[F_n, A_k]\| < 2^{-n}, \quad \|[F_n, \delta(A_k)]\| < 2^{-n},$$

for $1 \le k \le n$. Define

$$T = \sum F_n T_n F_n$$

and note from Theorem 4.1 that $T \in M(\mathfrak{A})$ with $0 \le T \le \|\delta\| I$. We have

$$[T, A_k] = \sum F_n T_n [F_n, A_k] + \sum F_n [T_n, A_k] F_n$$
$$+ \sum [F_n, A_k] T_n F_n,$$

and it follows from (iii) that the first and the third of these sums converge in $\mathfrak{A}$. For the middle terms we use the fact that $\sum F_n^2 = I$ (strict convergence) to compute

$$\sum F_n [T_n, A_k] F_n - \delta_0(A_k)$$
$$= \sum F_n([T_n, A_k] - \delta_0(A_k)) F_n - \sum [F_n, \delta_0(A_k)] F_n \in \mathfrak{A},$$

because $\|[T_n, A_k] - \delta_0(A_k)\| < 2^{-n+2}$ for $k \leq n$. Taken together it means that $[T, A_k] - \delta_0(A_k) \in \mathfrak{A}$ for every A_k in $\mathfrak{D}$. Now let $S = \pi(T)$, and it is immediate that $[S, B] = \delta(B)$ for every B in $\mathfrak{B}$, as desired.

$\square$

Corollary 10.2. *If δ is a derivation of a corona algebra $C(\mathfrak{A})$, there is for each separable C^*-subalgebra $\mathfrak{B}$ of $C(\mathfrak{A})$ an element T in $C(\mathfrak{A})$, such that $\delta(B) = [T, B]$ for every B in $\mathfrak{B}$.*

Proof: Let $\mathfrak{B}_1$ be the (separable) C^*-algebra generated by $\bigcup \delta^n(\mathfrak{B})$. Then $\mathfrak{B}_1$ is δ-invariant, so that $\delta | \mathfrak{B}_1$ is a derivation of $\mathfrak{B}_1$ and the previous result applies.

$\square$

Corollary 10.3. *Let $\mathfrak{I}$ be an essential ideal in a separable C^*-algebra $\mathfrak{A}$ (so that we have a canonical embedding $\mathfrak{I} \subset \mathfrak{A} \subset M(\mathfrak{I})$). There is then for each derivation δ of $\mathfrak{A}$ an element S in $M(\mathfrak{I})$ such that $\delta(A) - [S, A] \in \mathfrak{I}$ for every A in $\mathfrak{A}$.*

Remark 10.4. The result above is not the only attempt at localizing derivations modulo ideals. In [19] it was shown that for any derivation δ of a C^*-algebra $\mathfrak{A}$ and every $\varepsilon > 0$ there is an essential ideal $\mathfrak{I}$ of $\mathfrak{A}$ and an element S in $M(\mathfrak{I})$ such that $\|\delta(A) - [S, A]\| \leq \varepsilon \|A\|$ for every A in $\mathfrak{I}$. Despite the similarity in the vocabulary, the two results seems to have little in common.

Since orthogonal elements in quotient C^*-algebras can be lifted to orthogonal elements in the original algebras, it is quite easy to show that any quotient image of an SAW^*-algebra is again an SAW^*-algebra. By contrast, a C^*-subalgebra, even an hereditary one, of an SAW^*-algebra, need not be an SAW^*-algebra. In the case where $\mathfrak{A}$ is a commutative C^*-subalgebra, and thus of the form $C(X)$ for some sub-Stonean space X, a closed ideal $\mathfrak{I}$ of $\mathfrak{A}$ will be an SAW^*-algebra precisely when the corresponding open subset Y of X (for which $\mathfrak{I} = C_0(Y)$) has a basically isolated complement. This, by definition, means that no σ-compact, open subset of Y has any limit points in $X \setminus Y$. This condition translates quite easily into a non-commutative language via the notion of annihilators.

Proposition 10.5. *If $\mathfrak{I}$ is an hereditary C^*-subalgebra of an SAW^*-algebra $\mathfrak{A}$, then $\mathfrak{I}$ is itself an SAW^*-algebra if and only if each element in $\mathfrak{I}_+$ has a local unit in $\mathfrak{I}_+$. In particular, if $\mathfrak{I}$ is also an ideal, it will be an SAW^*-ideal if and only if $\mathfrak{B}^\perp + \mathfrak{I} = \mathfrak{A}$ for every hereditary σ-unital C^*-subalgebra $\mathfrak{B}$ of $\mathfrak{I}$.*

Proof: The existence of local units is clearly a necessary condition for $\mathfrak{I}$ to be an SAW^*-algebra. On the other hand, if it is satisfied, and if S, T is an orthogonal pair in $\mathfrak{I}_+$, there is an element R in the SAW^*-algebra $\mathfrak{A}$ such that $SR = S$ and $RT = 0$. Now use the condition to obtain an element E in $\mathfrak{I}_+$ which is a unit for $S + T$. Replacing R by ERE, which belongs to $\mathfrak{I}_+$, it is clear that we have a unit for S which annihilates T.

If now $\mathfrak{I}$ is an ideal, and $\mathfrak{B}$ is an hereditary, σ-unital C^*-subalgebra of $\mathfrak{I}$, there is a unit E for $\mathfrak{B}$ in $\mathfrak{I}_+$, because $\mathfrak{B}$ is singly generated as an hereditary algebra. Assuming, as we may, that $0 \leq E \leq I$ it follows that $(I - E)\mathfrak{A}(I - E) \subset \mathfrak{B}^\perp$, whence

$$\mathfrak{A} = (I - E + E)\mathfrak{A}(I - E + E) \subset \mathfrak{B}^\perp + \mathfrak{I}.$$

Conversely, if we always have $\mathfrak{A} = \mathfrak{B}^\perp + \mathfrak{I}$, take an element T in $\mathfrak{I}_+$ and let $\mathfrak{B} = (T\mathfrak{A}T)^=$ be the associated hereditary C^*-subalgebra. Let E be a unit for T in $\mathfrak{A}_+$, and by assumption write it in the form $E = S + R$ with S in $\mathfrak{I}$ and R in $\mathfrak{B}^\perp$. Then both S and S^* are units for T, so that S^*S is a unit for T in $\mathfrak{I}_+$, whence $\mathfrak{I}$ is an SAW^*-ideal by the first part of the proof.

$\square$

Proposition 10.6. *Let ρ be a morphism of an SAW^*-algebra $\mathfrak{A}$ and let $\mathfrak{I} = \ker \rho$. Then $\mathfrak{I}$ is an SAW^*-ideal of $\mathfrak{A}$ if and only if*

$$\rho(\mathfrak{B})^\perp = \rho(\mathfrak{B}^\perp)$$

for every hereditary, σ-unital C^*-subalgebra $\mathfrak{B}$ of $\mathfrak{A}$.

Proof: If the condition is satisfied, take $\mathfrak{B}$ inside $\mathfrak{I}$. Then $\rho(\mathfrak{B}) = 0$, so the condition asserts that $\rho(\mathfrak{B}^\perp) = \rho(\mathfrak{A})$, or, $\mathfrak{A} = \mathfrak{B}^\perp + \mathfrak{I}$. By Proposition 10.5, $\mathfrak{I}$ is therefore an SAW^*-ideal.

Conversely, if $\mathfrak{I}$ is an SAW^*-ideal and $\mathfrak{B} \subset \mathfrak{A}$, it suffices to show that each element T in $\mathfrak{A}_+$ such that $T\mathfrak{B} \subset \mathfrak{I}$ (i.e. $\rho(T) \in \rho(\mathfrak{B})^\perp$) can be written in the form $T = S + R$, where $S \in \mathfrak{B}^\perp$ and $R \in \mathfrak{I}$. Towards this end, let H be a strictly positive element in $\mathfrak{B}_+$. Since $T\mathfrak{B} \subset \mathfrak{I}$, the same is true for any continuous function of T (vanishing at zero), in particular $T^{1/4}\mathfrak{B} \subset \mathfrak{I}$. Let $\mathfrak{D}$ be the (σ-unital) hereditary C^*-subalgebra of $\mathfrak{A}$ generated by $T^{1/4}H^2T^{1/4}$. Then $\mathfrak{D} \subset \mathfrak{I}$, so that $\mathfrak{D}^\perp + \mathfrak{I} = \mathfrak{A}$ by 10.5. In particular, $T^{1/4} = A + B$ for some A in $\mathfrak{D}^\perp$ and B in $\mathfrak{I}$. Put $S = T^{1/4}A^*AT^{1/4}$, and note that $T - S \in \mathfrak{I}$. Furthermore,

$$SH = T^{1/4}A^*(AT^{1/4}H) = 0$$

so that $S \in \mathfrak{B}^\perp$.

$\square$

Lemma 10.7. *If $\mathfrak{I}$ is an ideal in a σ-unital C^*-algebra $\mathfrak{A}$, and $T \in M(\mathfrak{A})_+$, such that $T\mathfrak{A} \subset \mathfrak{I}$, then there is an element E in $M(\mathfrak{A})$, with $0 \leq E \leq I$ such that $E\mathfrak{A} \subset \mathfrak{I}$ and $T(I - E) \in \mathfrak{I}$.*

Proof: Take a strictly positive element H in $\mathfrak{A}$, and let $\mathfrak{D}$ denote the separable C^*-subalgebra of $M(\mathfrak{A})$ generated by T and H. Put

$$\mathfrak{J}_1 = \mathfrak{I} \cap \mathfrak{D}, \quad \mathfrak{J}_2 = \mathfrak{J}_1^{\perp} \cap \mathfrak{D}, \quad \mathfrak{J} = \mathfrak{J}_1 + \mathfrak{J}_2.$$

Then $\mathfrak{J}$ is a σ-unital C^*-algebra and an essential ideal in $\mathfrak{D}$, so that we have a natural embedding

$$\mathfrak{J} \subset \mathfrak{D} \subset M(\mathfrak{J}),$$

cf. Theorem 3.1. Since $TH \in \mathfrak{J}_1$ and $C(\mathfrak{J})$ is an SAW^*-algebra, there is an element E in $M(\mathfrak{J})$ with $0 \leq E \leq I$ such that

$$T(I - E) \in \mathfrak{J}, \quad EH \in \mathfrak{J}.$$

The direct sum $\mathfrak{J} = \mathfrak{J}_1 + \mathfrak{J}_2$ implies that we have a similar direct sum $M(\mathfrak{J}) = M(\mathfrak{J}_1) + M(\mathfrak{J}_2)$, so that there is an orthogonal decomposition $E = E_1 + E_2$. Realizing $M(\mathfrak{A})$ concretely as the idealizer of $\mathfrak{A}$ in the von Neumann algebra $\mathfrak{A}^{**}$, see p. 9, we see that since $\mathfrak{J} \subset \mathfrak{A}^{**}$, $M(\mathfrak{J}) \subset \mathfrak{A}^{**}$. In this algebra $H\mathfrak{A}$ is weakly dense, so the observation that $E_2 TH \in \mathfrak{J}_2 \cap \mathfrak{J}_1 = \{0\}$, implies that $E_2 T = 0$. Now $E_1 H \in \mathfrak{J}_1$ which shows that

$$E_1 \mathfrak{A} \in (E_1 H \mathfrak{A})^{=} \subset (\mathfrak{J}_1 \mathfrak{A})^{=} \subset \mathfrak{I}.$$

Finally, to prove that $T(I - E_1) \in \mathfrak{I}$, note that if $T = T_1 + T_2$ is the orthogonal decomposition in $M(\mathfrak{J}_1) + M(\mathfrak{J}_2)$, then

$$T_2 H = TH - T_1 H \in \mathfrak{J}_1 \cap M(\mathfrak{J}_2) = \{0\}$$

whence $T_2 = 0$, arguing as above. Thus

$$T(I - E_1) \in M(\mathfrak{J}_1) \cap \mathfrak{J} = \mathfrak{J}_1 \subset \mathfrak{I},$$

as desired.

$\square$

Let $\rho : \mathfrak{A} \to \mathfrak{B}$ be a morphism between σ-unital C^*-algebras $\mathfrak{A}$ and $\mathfrak{B}$. Then as discussed in the proof of Theorem 4.3 there is a canonical weakly continuous extension $\rho^{**} : \mathfrak{A}^{**} \to \mathfrak{B}^{**}$. It is an elementary computation to show that $\rho^{**}(M(\mathfrak{A})) \subset M(\mathfrak{B})$, provided that $\rho(\mathfrak{A})$ contains an approximate unit for $\mathfrak{B}$.

Theorem 10.8. *Let* $\rho : \mathfrak{A} \to \mathfrak{B}$ *be a morphism between* σ-*unital* C^*-*algebras such that* $\rho(\mathfrak{A})$ *contains an approximate unit for* $\mathfrak{B}$. *Then the canonical extension* $\rho^{**} : M(\mathfrak{A}) \to M(\mathfrak{B})$ *induces a morphism* $\tilde{\rho} : C(\mathfrak{A}) \to C(\mathfrak{B})$ *between the corona algebras such that* $\ker \tilde{\rho}$ *is an* SAW^*-*ideal of* $C(\mathfrak{A})$, *isomorphic to* $M(\mathfrak{B}, \ker \rho)/\ker \rho$, *where*

$$M(\mathfrak{A}, \ker \rho) = \{T \in M(\mathfrak{A}) \mid T\mathfrak{A} \subset \ker \rho\}.$$

Proof: Consider the diagram

$$
\begin{array}{ccccc}
\ker \rho & \overset{\iota}{\longrightarrow} & \mathfrak{A} & \overset{\rho}{\longrightarrow} & \mathfrak{B} \\
\iota \downarrow & & \iota \downarrow & & \iota \downarrow \\
\ker \rho^{**} & \overset{\iota}{\longrightarrow} & M(\mathfrak{A}) & \overset{\rho^{**}}{\longrightarrow} & M(\mathfrak{B}) \\
\sigma \downarrow & & \pi \downarrow & & \pi \downarrow \\
\ker \tilde{\rho} & \overset{\iota}{\longrightarrow} & C(\mathfrak{A}) & \overset{\tilde{\rho}}{\longrightarrow} & C(\mathfrak{B})
\end{array}
$$

Here the maps denoted by ι are the obvious embeddings, and those denoted by π are the quotient maps. Clearly

$$\ker \rho^{**} = \{T \in M(\mathfrak{A}) \mid T\mathfrak{A} \subset \ker \rho\} = M(\mathfrak{A}, \ker \rho);$$

in particular, the upper half of the diagram is commutative. The map $\tilde{\rho}$ is defined by

$$\tilde{\rho}(\pi(T)) = \pi(\rho^{**}(T))$$

for each T in $M(\mathfrak{A})$. Since $\pi(T) = 0$ only if $T \in \mathfrak{A}$, in which case $\rho^{**}(T) \in \mathfrak{B}$, this is an admissible definition, and produces a morphism from $C(\mathfrak{A})$ into $C(\mathfrak{B})$. We see that

$$\pi^{-1}(\ker \tilde{\rho}) = \{T \in M(\mathfrak{A}) \mid \rho^{**}(T) \in \mathfrak{B}\}.$$

But in that case if (E_n) is an approximate unit for $\mathfrak{A}$, then $(\rho(E_n))$ is an approximate unit for $\mathfrak{B}$, whence $\rho^{**}(T(I - E_n)) \to 0$. Thus $T \in \mathfrak{A} + \ker \rho^{**}$, i.e.

$$\pi^{-1}(\ker \tilde{\rho}) = \mathfrak{A} + M(\mathfrak{A}, \ker \rho).$$

We can therefore unambiguously define the map σ as $\iota^{-1} \circ \pi \circ \iota$, and it follows that σ is a surjective quotient map which makes the whole diagram commutative. Furthermore

$$\ker \tilde{\rho} = \pi(\mathfrak{A} + M(\mathfrak{A}, \ker \rho))$$

$$= M(\mathfrak{A}, \ker \rho)/M(\mathfrak{A}, \ker \rho) \cap \mathfrak{A} = M(\mathfrak{A}, \ker \rho)/\ker \rho.$$

To show that $\ker \tilde{\rho}$ is an SAW^*-ideal, it suffices by Proposition 10.5 to show that every element in $\ker \tilde{\rho}$ has a local unit. Translated upstairs to preimages it means that for each positive element T in $M(\mathfrak{A}, \ker \rho)$ $(= \sigma^{-1}(\ker \tilde{\rho}))$ we must find an element E in $M(\mathfrak{A}, \ker \rho)$, with $0 \le E \le I$, such that $T(I - E) \in \ker \rho$. But that is precisely the contents of Lemma 10.7.

$\square$

Corollary 10.9. *If $\rho : \mathfrak{A} \to \mathfrak{B}$ is a surjective morphism between σ-unital C^*-algebras, and $\tilde{\rho} : C(\mathfrak{A}) \to C(\mathfrak{B})$ is the induced surjective morphism between the corona algebras, then*

$$\tilde{\rho}(\mathfrak{D})^{\perp} = \tilde{\rho}(\mathfrak{D}^{\perp})$$

for every hereditary, σ-unital C^*-subalgebra $\mathfrak{D}$ of $C(\mathfrak{A})$.

Proof: Combine 10.8 and 10.6.

$\square$

Theorem 10.10. *If $\rho : \mathfrak{A} \to \mathfrak{B}$ is a surjective morphism between σ-unital C^*-algebras, and $\tilde{\rho} : C(\mathfrak{A}) \to C(\mathfrak{B})$ is the induced surjective morphism between the corona algebras, then*

$$\tilde{\rho}(\mathfrak{D})' = \tilde{\rho}(\mathfrak{D}')$$

for every separable C^*-subalgebra $\mathfrak{D}$ of $C(\mathfrak{A})$ (where $'$ denotes relative commutants).

Proof: Clearly $\tilde{\rho}(\mathfrak{D}') \subset \tilde{\rho}(\mathfrak{D})'$, so it suffices to show that every T in $C(\mathfrak{A})_{sa}$, such that $\tilde{\rho}(T) \in \tilde{\rho}(\mathfrak{D})'$, can be perturbed inside $\ker \tilde{\rho}$ to an element in $\mathfrak{D}'$.

Let (D_n) be a countable, dense, $*$-invariant subring in $\mathfrak{D}$, and consider the hereditary, σ-unital C^*-subalgebra $\mathfrak{C}$ of $C(\mathfrak{A})$ generated by the sequence $([D_n, T])$ of commutators. By assumption $\mathfrak{C} \subset \ker \tilde{\rho}$. Now $\mathfrak{C} = \mathfrak{L} \cap \mathfrak{L}^*$ for some left ideal $\mathfrak{L}$ in $\ker \tilde{\rho}$, and replacing if necessary $\mathfrak{L}$ by $(\bigcup \mathfrak{L}D_n)^=$ (taking $D_0 = I$) we may assume that $\mathfrak{L}\mathfrak{D} \subset \mathfrak{L}$, which implies that $\mathfrak{D}\mathfrak{C} \subset \mathfrak{C}$.

Since $\mathfrak{C}$ is σ-unital, it has a strictly positive element H. However, $H \in \ker \tilde{\rho}$ which by Theorem 10.8 is an SAW-ideal. Thus by Proposition 10.5 there is an element R in $\ker \tilde{\rho}$ with $0 \leq R \leq I$ such that $RH = H$. It follows that $(I - R)\mathfrak{C} = 0$ and thus $(I - R)\mathfrak{D}\mathfrak{C} = 0$ because $\mathfrak{D}\mathfrak{C} \subset \mathfrak{C}$. In particular $(I - R)\mathfrak{D}H = 0$. By Theorem 8.3 there is therefore an element S in $C(\mathfrak{A})$, $0 \leq S \leq I$, commuting with $\mathfrak{D}$, such that $(I - R)(I - S) = 0$ and $SH = 0$. This means that $S\mathfrak{C} = 0$ and in particular $S[T, D] = 0$ for every D in $\mathfrak{D}$. Since $S \in \mathfrak{D}'$ we conclude that

$$[STS, D] = S[T, D]S = 0,$$

so that $STS \in \mathfrak{D}'$. Finally

$$T - STS = (I - S)T + ST(I - S) \in \ker \tilde{\rho},$$

84

because $I - S = R(I - S) \in \ker \tilde{\rho}$.

$\square$

Remark 10.11. Noting the parallel between 10.9. and 10.10., the reader might feel that also Theorem 7.7 about double annihilators should have a commutator version. Ideally the statement should be that for every separable C^*-subalgebra $\mathfrak{D}$ of a corona algebra $C(\mathfrak{A})$ we have

$$\mathfrak{D}'' = \mathfrak{D}.$$

The statement is blatantly false, in general. If for example $\mathfrak{A}$ is commutative, then $\mathfrak{D}' = \mathfrak{D}'' = C(\mathfrak{A})$ for any subalgebra $\mathfrak{D}$ of $C(\mathfrak{A})$. The tantalizing fact is that by Voiculescu's result [29] the statement holds if $\mathfrak{A} = \mathfrak{K}$. The challenge is therefore to find suitable extra conditions on $\mathfrak{A}$ that will assure the validity of the double commutator theorem. One first guess would be to assume that $\mathfrak{A}$ was a maximal ideal in $M(\mathfrak{A})$, i.e. that $C(\mathfrak{A})$ is simple; but whether this condition is enough to give the ultra-high degree of non-commutativity that we need is not known.

§11. A Lifting Problem

Let $\rho : \mathfrak{A} \to \mathfrak{B}$ be a surjective morphism between C^*-algebras. We consider a property (p) concerning elements in C^*-algebras and denote by $\mathfrak{A}_p$ the class of elements in $\mathfrak{A}$ that has the property (p). Each (p) gives a lifting problem:

$$\text{Is } \rho(\mathfrak{A}_p) = \mathfrak{B}_p?$$

Note that since ρ is structure preserving we may always assume that $\rho(\mathfrak{A}_p) \subset \mathfrak{B}_p$. Consider the following list

property	answer
Being self-adjoint	yes
Being positive	yes
Being unitary	no
Being normal	no
Being a projection	no
Being of norm one	yes
Being a positive orthogonal sequence	yes
Having spectral radius 1	yes
Having spectral radius 0	?

Having square zero	yes
Being nilpotent of order n	yes!

The last four lifting problems were first studied for $\mathfrak{A} = B(\mathfrak{H})$ (so $\mathfrak{B}$ is the Calkin algebra), and being solved there, the same methods would then give the result for an arbitrary von Neumann algebra $\mathfrak{A}$, see [17] and [2].

The problem of lifting an element with a fixed, non-zero, spectral radius was solved in [2] for general C^*-algebras, and in the same paper is was shown that any element with square zero could be lifted to a similar element. With the aid of the SAW^*-property in corona algebras we shall now prove that any nilpotent element can be lifted. We warm up by showing a simpler result, also mentioned in [2].

Lemma 11.1. *If $\mathfrak{A}$ is a σ-unital C^*-algebra and $T_1, \ldots, T_n$ are elements in $M(\mathfrak{A})$ such that $\Pi T_k \in \mathfrak{A}$, then there are elements $A_1, \ldots, A_n$ in $\mathfrak{A}$ such that the product $\Pi(T_k - A_k) = 0$.*

Proof: Let π denote the quotient map onto the corona algebra, and assume that for some n in $\mathbf{N}$ the lemma has been established for all $k \leq n$. If now $T_1, \ldots, T_{n+1}$ are elements in $M(\mathfrak{A})$ such that $\Pi T_k \in \mathfrak{A}$, let $S_n = \Pi_{k=1}^n T_k$. Then $\pi(S_n)\pi(T_{n+1}) = 0$, and since $C(\mathfrak{A})$ is an SAW^*-algebra by Corollary 7.5, there are elements R and Q in $M(\mathfrak{A})$, $0 \leq R, Q \leq I$, such that

$$(*) \qquad S_n(I - R) \in \mathfrak{A}, (I - Q)T_{n+1} \in \mathfrak{A}, RQ \in \mathfrak{A}.$$

Replacing R and Q with $(R - Q)_+$ and $(R - Q)_-$ we may assume that $RQ = 0$.

The induction hypothesis, applied to the n elements $T_1, \ldots, T_{n-1}, T_n(I - R)$, whose product lies in $\mathfrak{A}$ by $(*)$, produces elements $A_1, \ldots, A_n$ in $\mathfrak{A}$ such that with $S_{n-1} = \Pi_{k=1}^{n-1}(T_k - A_k)$ we have $S_{n-1}(T_n(I - R) - A_n) = 0$. Put $A_{n+1} = (I - Q)T_{n+1}$, and note from $(*)$ that $A_{n+1} \in \mathfrak{A}$. Finally,

$$\prod_{k=1}^{n+1}(T_k - A_k) = S_{n-1}(T_n - A_n)(T_{n+1} - A_{n+1})$$
$$= S_{n-1}(T_n(I - R) - A_n + T_n R)QT_{n+1} = 0 + 0 = 0.$$

The proof is completed by induction.

$\square$

Theorem 11.2. *If $\mathfrak{I}$ is a closed ideal in a C^*-algebra $\mathfrak{A}$, and if $T_1, \ldots, T_n$ are elements in $\mathfrak{A}$ such that $\Pi T_k \in \mathfrak{I}$, then there are elements $A_1, \ldots, A_n$ in $\mathfrak{I}$ such that $\Pi(T_k - A_k) = 0$.*

Proof: We may assume that $\mathfrak{A}$ is separable, replacing if necessary $\mathfrak{A}$ with the C^*-algebra $\mathfrak{A}_0$ generated by $T_1, \ldots, T_n$ and $\mathfrak{I}$ with $\mathfrak{I} \cap \mathfrak{A}_0$. Let now $\mathfrak{I}^\perp$ denote the annihilator of $\mathfrak{I}$ in $\mathfrak{A}$. Then $\mathfrak{I}^\perp$ is a closed ideal in $\mathfrak{A}$, orthogonal to $\mathfrak{I}$ and $\mathfrak{J} = \mathfrak{I} + \mathfrak{I}^\perp$ is an essential ideal in $\mathfrak{A}$. By Theorem 3.1 there is therefore a natural embedding $\mathfrak{J} \subset \mathfrak{A} \subset M(\mathfrak{J})$. Applying Lemma 11.1. to the separable, hence σ-unital C^*-algebra $\mathfrak{J}$, we obtain elements $C_1, \ldots, C_n$ in $\mathfrak{J}$ such that $\Pi(T_k - C_k) = 0$. Since $\mathfrak{I} \cap \mathfrak{I}^\perp = 0$, each C_k has a unique decomposition $C_k = A_k + B_k$. Thus

$$0 = \Pi(T_k - A_k - B_k) = \Pi(T_k - A_k) + B,$$

where the element B is the sum of products, each containing a factor B_k. Thus $B \in \mathfrak{I}^\perp$, whereas $\Pi(T_k - A_k) \in \mathfrak{I}$ by assumption. Since $\mathfrak{I} \cap \mathfrak{I}^\perp = 0$ it follows that $\Pi(T_k - A_k) = 0$ (and $B = 0$) as desired.

$\square$

Lemma 11.3. *If T is an element in a corona algebra $C(\mathfrak{A})$, such that $T^n = 0$ for some number n, then there are elements $R_0, R_1, \ldots, R_n$ in $C(\mathfrak{A})$, $0 \leq R_k \leq I$, such that*

(i) $R_k R_{k-1} = R_{k-1}$
(ii) $(I - R_k)T^{n-k} = 0$
(iii) $(I - R_{k-1})TR_k = 0$

for $1 \leq k \leq n$, where $R_0 = 0$ and $R_n = I$ by definition.

Proof: Suppose that we have already constructed $R_0, R_1, \ldots, R_m$ in $C(\mathfrak{A})$, satisfying the conditions (i)–(iii) for all $k \leq m$, where $0 \leq m < n$. Then

$$(I - R_m)TT^{n-m-1} = 0$$

by (ii). Furthermore, by (i) and (iii)

$$(I - R_m)TR_m = (I - R_m)R_{m-1}TR_m = 0,$$

so that

$$((I - R_m)T)(T^{n-m-1}(T^{n-m-1})^* + R_m) = 0.$$

Since $C(\mathfrak{A})$ is an SAW^*-algebra by Corollary 7.5, there is an element R_{m+1}, with $0 \leq R_{m+1} \leq I$, such that

$$(I - R_m)TR_{m+1} = 0,$$

$$(I - R_{m+1})T^{n-m-1} = (I - R_{m+1})R_m = 0.$$

Thus $R_0, \ldots, R_{m+1}$ satisfy (i)–(iii), and the proof is completed by induction.

$\square$

Remark 11.4. If $T^n = 0$ for an element T in $\mathbb{B}(\mathfrak{H})$ there is a fairly canonical way of constructing elements $R_0, \ldots, R_n$ satisfying the conditions (i)–(iii) in 11.2. Simply let R_k be the range projection of T^{n-k}. In a corona setting the R_k's are not necessarily projections, and there is no canonical choice for them. Nevertheless they have the same purpose of decomposing T in a triangular form relative to a commutative algebra, and thus prepare the way for a lifting.

Lemma 11.5. *If $\rho : \mathfrak{A} \to \mathfrak{B}$ is a surjective morphism between unital C^*-algebras and (R_n) is a sequence in $\mathfrak{B}$, with $0 \leq R_n \leq I$ and $R_n R_{n+1} = R_n$ for all n, then there is a similar sequence (Q_n) in $\mathfrak{A}$ with $\rho(Q_n) = R_n$ for all n.*

Proof: Suppose for some number n we have chosen $Q_1, \ldots, Q_{n-1}$ in $\mathfrak{A}$ and an hereditary C^*-subalgebra $\mathfrak{A}_n$ of $\mathfrak{A}$, orthogonal to Q_{n-1}, such that

(i)
$$0 \leq Q_k \leq I, \quad Q_{k-1}Q_k = Q_{k-1}, \quad \rho(Q_k) = R_k$$

for $1 \leq k < n$, and

(ii)
$$\{I - R_m \mid m \geq n\} \subset \rho(\mathfrak{A}_n).$$

Note that for $n = 1$ we have chosen no Q's and we may take $\mathfrak{A}_1 = \mathfrak{A}$. Put $\widetilde{\mathfrak{A}}_n = \mathfrak{A}_n + \mathbb{C}I$ and $\widetilde{\mathfrak{B}}_n = \rho(\widetilde{\mathfrak{A}}_n)$. As $R_n(I - R_{n+1}) = 0$ in $\widetilde{\mathfrak{B}}_n$, there are orthogonal, positive elements Q_n and S in $\widetilde{\mathfrak{A}}_n$, such that $\rho(Q_n) = R_n$ and $\rho(S) = I - R_{n+1}$. Simply take $Q_n = A_+$, $S = A_-$, where A is a self-adjoint preimage of $R_n - (I - R_{n+1})$ in $\widetilde{\mathfrak{A}}_n$. Let $\mathfrak{A}_{n+1}$ be the hereditary C^*-subalgebra of $\mathfrak{A}$ generated by S, i.e. $\mathfrak{A}_{n+1} = (S\mathfrak{A}S)^=$. Clearly, then, $\mathfrak{A}_{n+1}$ is orthogonal to Q_n. Moreover, for $m \geq n + 1$,

$$I - R_m \in (I - R_{n+1})\mathfrak{B}(I - R_{n+1}) \subset \rho(\mathfrak{A}_{n+1}).$$

Finally, if $n > 1$, $\rho(\mathfrak{A}_n)$ is a proper, hereditary subalgebra of $\mathfrak{B}$. Since $Q_n = \lambda I + A$ with A in $\mathfrak{A}_n$ and

$$\rho(I - Q_n) = I - \lambda I - \rho(A) \in \rho(\mathfrak{A}_n),$$

it follows that $\lambda = 1$. Consequently,

$$Q_{n-1}Q_n = Q_{n-1}(I + A) = Q_{n-1},$$

because $Q_{n-1}\mathfrak{A}_n = 0$, so that the pair $Q_n, \mathfrak{A}_{n+1}$ satisfy the requirements in (i) and (ii). The sequence (Q_n) can therefore be constructed by induction.

$\square$

Lemma 11.6. *If T is an element of a corona algebra $C(\mathfrak{A})$, such that $T^n = 0$, then there is an element S in $M(\mathfrak{A})$ with $S^n = 0$ such that $\pi(S) = T$, where π denotes the quotient map.*

Proof: Choose by Lemma 11.3 elements $R_0, \ldots, R_n$ satisfying (i), (ii) and (iii) in that lemma. Then use Lemma 11.5 to lift the R's to be a set $\{Q_0, \ldots, Q_n\}$ with $Q_0 = 0$, $Q_n = I$, $0 \leq Q_k \leq I$, and $Q_k Q_{k-1} = Q_{k-1}$ for all $k \leq n$.

Now take continuous functions f and g on $[0, 1]$ with $f(0) = g(0) = 0$, $f(1) = g(1) = 1$, and $fg = f$. Then let B be an element in $M(\mathfrak{A})$ with $\pi(B) = T$ and define

$$S_k = f(Q_{k-1})B(g(Q_k) - g(Q_{k-1}))$$

for $1 \leq k \leq n$. Note that $S_1 = 0$ (as $Q_0 = 0$) and that

$$\pi(S_k) = f(R_{k-1})T(g(R_k) - g(R_{k-1})) = T(g(R_k) - g(R_{k-1})),$$

since $(I - f(R_{k-1}))Tg(R_k) = 0$ by (iii), because g can be approximated by a polynomial in the variable t, and $1 - f$ by a polynomial in $1 - t$. Thus, with $S = \sum S_k$, we have

$$\pi(S) = \sum_{k=1}^{n} T(g(R_k) - g(R_{k-1}))$$
$$= T(g(R_n) - g(R_0)) = T.$$

Since $(g(Q_k) - g(Q_{k-1}))f(Q_{j-1}) = 0$ for $j \leq k$, we see from the construction of the S_k's that $S_k S_j = 0$ if $j \leq k$. Now S^n is the sum of products, each of which must contain n factors from the set $\{S_2, \ldots, S_n\}$. Each must therefore have the form $S_{\alpha_1} S_{\alpha_2} \ldots S_{\alpha_n}$ with $\alpha_k \leq \alpha_{k+1}$ for some k, and therefore be zero. Thus $S^n = 0$, as desired.

$\square$

Theorem 11.7. *If $\mathfrak{J}$ is a closed ideal in a C^*-algebra $\mathfrak{A}$, and $T \in \mathfrak{A}$ such that $T^n \in \mathfrak{J}$, then $(T - A)^n = 0$ for some A in $\mathfrak{J}$.*

Proof: As in Theorem 11.2 we may assume that $\mathfrak{A}$ is separable, passing if necessary to the C^*-algebra generated by T. Likewise, we let $\mathfrak{J}^{\perp}$ be the annihilator of $\mathfrak{J}$ in $\mathfrak{A}$, and put $\mathfrak{I} = \mathfrak{J} + \mathfrak{J}^{\perp}$; to obtain an embedding

$$\mathfrak{I} \subset \mathfrak{A} \subset M(\mathfrak{I}).$$

Applying Lemma 11.6 to the σ-unital algebra $\mathfrak{I}$, we find an element $C = A + B$ in $\mathfrak{I} = \mathfrak{J} + \mathfrak{J}^{\perp}$ such that

$$0 = (T - C)^n = (T - A)^n + R,$$

where R is a sum of products, each containing a factor B. Thus $R \in \mathfrak{J}^{\perp}$, and since $(T - A)^n \in \mathfrak{J}$ by assumption, it follows that $(T - A)^n = 0$ (and $R = 0$).

$\square$

References

1. Akemann, C. A., The general Stone-Weierstrass problem, J. Functional Analysis **4** (1969), 277–294.

2. Akemann, C. A. and G. K. Pedersen, Ideal perturbations of elements in C^*-algebras, Math. Scand. **41** (1977), 117–139.

3. Akemann,C. A., G. K. Pedersen and J. Tomiyama, Multipliers of C^*-algebras, J. Functional Analysis **13** (1973), 277-301.

4. Arveson, W. B., Notes on extensions of C^*-algebras, Duke Math. J. **44** (1977), 329–35.

5. Berberian, S., $N \times N$ matrices over an AW^*-algebra, Amer. J. Math. **80** (1958), 37–44.

6. Berberian, S., Baer *-Rings. Ergebn. d. Math. u.i. Grenzgeb., Springer-Verlag, 1972.

7. Blackadar, B., *K-Theory for Operator Algebras*, MSRI publ. **5**, Springer-Verlag, 1986.

8. Busby, R. C., Double centralizers and extensions of C^*-algebras, Trans. Amer. Math. Soc. **132** (1968), 79–99.

9. Combes, F., Sur les faces d'une C^*-algèbre, Bull. Sci. Math. **93** (1969), 37–62.

10. Cuntz, J. and N. Higson, Kuiper's theorem for Hilbert modules, Contemporary Math. **62** (1978), 429–433.

11. Effros, E. G., Order ideals in a C^*-algebra and its dual, Duke Math. J. **30** (1963), 391–412.

12. Grove, K. and G. K. Pedersen, Sub-Stonean spaces and corona sets, Functional Analysis **56** (1984), 124–143.

13. Herman, R. H. and L. N. Vaserstein, The stable rangle of C^*-algebras, Inventiones Math.**77** (1984), 553–555.

14. Kasparov, G. G., The operator K-functor and extensions of C^*-algebras, Math. USSR Izv.**16** (1981), 513–572.

15. Mingo, J., K-theory and multipliers of stable C^*-algebras, Trans. Amer. Math. Soc. **299** (1987), 397–412.

16. Olesen, D. and G. K. Pedersen, Applications of the Connes spectrum to C^*-dynamical systems, III, J. Functional Analysis **45** (1982), 357–390.

17. Olsen, C. L., A structure theorem for polynomially compact operators, Amer. J. Math. **93** (1971), 686–698.

18. Olsen, C. L. and G. K. Pedersen, Corona C^*-algebras and their applications to lifting problems, Math. Scand., to appear.

19. Pedersen, G. K., Approximating derivations on ideals of C^*-algebras, Inventiones Math. **45** (1978), 299-305.

20. Pedersen, G. K., C^*-algebras and their Automorphism Groups, L.M.S. monogr. **14**, Academic Press, 1979.

21. Pedersen, G. K., SAW^*-algebras and corona C^*-algebras, contributions to non-commutative topology, J. Operator Theory **15** (1986), 15–32.

22. Pedersen, G. K., Unitary extensions and polar decompositions in a C^*-algebra, J. Operator Theory **17** (1987), 357–364.

23. Pedersen, G. K., Three quavers on unitary elements in C^*-algebras, Pac. J. Math. **137** (1989), 169–179.

24. Rieffel, M. A., Dimensions and stable rank in the K-theory of C^*-algebras, Proc. London Math. Soc. **46** (1983), 301–333.

25. Robertson, A. G., Stable range in C^*-algebras, Proc. Camb. Phil. Soc. **87** (1980), 413–418.

26. Rørdam, M., Advances in the theory of unitary rank and regular approximation, Annals of Math. **128** (1988), 153–172.

27. Smith, R. R. and D. P. Williams, The decomposition property for C^*-algebras, J. Operator Theory **16** (1986), 51–74.

28. Smith, R. R. and D. P. Williams, Separable injectivity for C^*-algebras, Ind. Univ. Math. J. **37** (1988), 111-133.

29. Voiculescu, D. V., A non-commutative Weyl–von Neumann theorem, Rev. Roum. Math. **21** (1976), 97–113.

30. Woronowicz, S. L., A remark on the polar decomposition of M-sectorial operators, Lett. Math. Phys. **1** (1977), 429–433.

31. Zhang, S., A Riesz decomposition property and ideal structure of multiplier algebras, Preprint, 1988.

32. Zhang, S., K_1-groups, quasidiagonality and interpolation by multiplier projections, Preprint, 1988.

33. Zhang, S., On the structure of projections and ideals of corona algebras, Canad. J. Math., to appear.

34. Zhang, S., C^*-algebras with real rank zero and the internal structure of their corona and multiplier algebras. Part I, II, III and IV, Preprints, 1989.

Gert K. Pedersen

Københavns Universitets Matematiske Institut

Universitetsparken 5

2100 Københavns øDanmard

C. Peligrad

Crossed product C*-algebras by compact group actions vs. fixed point algebras

Let G be a locally compact group and α a strongly continuous action of G as *-automorphism of a C^*-algebra A. One can associate with this system the C^*-crossed product $G \times_\alpha A$ as follows: Consider in $L^1(G, A)$ the following operations

$$(y * z)(g_0) = \int_G y(g)\alpha_g \left(z\left(g^{-1}g_0\right)\right) dg$$

$$y^*(g) = \alpha_g \left(y\left(g^{-1}\right)^*\right) \Delta\left(g^{-1}\right)$$

$$\|y\| = \int |y(g)|dg$$

where $y, z \in L^1(G, A)$ and Δ is the modular function of G. Then $G \times_\alpha A$ is simply the enveloping C^*-algebra of $L^1(G, A)$ [5].

Let $K \subset G$ be a compact subgroup. Denote by $\widehat{K}$ its dual (i.e., the set of unitary equivalence classes of irreducible unitary representations). For each $\pi \in \widehat{K}$ we denote also by π a fixed representative of π. Let H_π be the Hilbert space of π and d_π the dimension of H_π. For every $\pi \in \widehat{K}, \chi_\pi$ denotes the normalized character of π, i.e., $\chi_\pi(k) = d_\pi \cdot tr(\pi_{k^{-1}}), k \in K$.

We denote by i the trivial one dimensional representation of K.

$L^1(K)$ can be embedded as follows in the multiplier algebra $M(G \times_\alpha A)$ of $G \times_\alpha A$:

$$(\phi * y)(g) = \int_k \phi(k)\alpha_k \left(y(k^{-1}g)\right) dk$$

$$(y * \phi)(g) = \int_k \phi(k)y(gk^{-1})dk$$

where $\phi \in L^1(K)$ and $y \in L^1(G, A)$. For $\pi_1, \pi_2 \in \widehat{K}$ let $S_{\pi_1, \pi_2} = \chi_{\pi_1} * (G \times_\alpha A) * \chi_{\pi_2}$. If $\pi_1 = \pi_2$ we get $S_\pi = S_{\pi, \pi}$. The subspaces S_{π_1, π_2} are the analogues of some subspaces of spherical functions considered in [3]. In the more general context of C^*-algebras these subspace have been considered for compact G in [4] and in the general case in [6]. If $G = K$ then $S_{\pi, i}$ is (linearly) isomorphic with $A_1^\alpha(\pi) = \{\int \chi_\pi(k)\alpha_k(a)dk | a \in A\}$, which in turn is (linearly) isomorphic with

$$A_2^\alpha(\pi) = \{[a_{ij}] \in A \otimes B(H_\pi) \,|\, [\alpha_k(a_{ij})] = [a_{ij}](| \otimes \pi_k), k \in K\}. \tag{6}$$

Therefore, the subspaces $S_{\pi,i}$ are the analogues of the spectral subspaces for compact group actions.

Similarly, the subalgebras S_π are the analogues of the fixed point algebras $A \otimes B(H_\pi)^{\alpha \otimes ad\pi}$ for compact groups [6].

In what follows we shall discuss certain properties which the crossed product $G \times_\alpha A$ may have, in terms of A and $S_\pi, \pi \in \widehat{K}$.

Consider the following:

Problem. What conditions about the group G and the action α do guarantee the equivalence of the following properties:

(1) $G \times_\alpha A$ is nuclear,

(2) S_i is nuclear,

(3) A is nuclear.

In particular, are the above conditions equivalent if $G = K$?

If $G = K$ the problem reduces to 2) $\Rightarrow$ 3). Indeed, if 2) $\Rightarrow$ 3), then by general properties of crossed products of nuclear C^*-algebras by amenable group actions it follows that $G \times_\alpha A$ is nuclear. Then, S_i can be identified with a hereditary C^*-subalgebra of $G \times_\alpha A$ [9] and therefore S_i is nuclear. In [8] we proved the following

Proposition 1. *The following conditions are equivalent:*

(i) *$G \times_\alpha A$ is nuclear,*

(ii) *S_π are nuclear, $\pi \in \widehat{K}$.*

If G is amenable, any of the above conditions is equivalent with A being nuclear.

In what follows we restrict to the case $G = K$. Then S_i is isomorphic with the fixed point algebra A^α. If α is ergodic, i.e., $A^\alpha = \mathbb{C}I$ then by [10] A is nuclear so that the answer to our Problem is affirmative. We observed in [8] that the answer to our Problem is affirmative also if A^α is an AF C^*-algebra.

Next, we shall consider other situations in which the nuclearity of A^α implies that of A.

The action α is called saturated if

$$A_2^\alpha(\pi)^* \overline{A_2^\alpha(\pi)} = A \otimes B(H_\pi)^{\alpha \otimes ad\pi}.$$

Proposition 2. *Let G be a compact separable group, acting on a separable C^*-algebra A. Assume that the action is saturated. Then, if A^α is nuclear it follows that A is nuclear.*

Proof: It is easy to show that $A_2^\alpha(\pi)$ is an $A_2^\alpha(\pi)^* \overline{A_2^\alpha(\pi)} - A_2^\alpha \overline{A_2^\alpha(\pi)^*}$ imprimitivity bimodule. Since A^α is nuclear, we have that $A^\alpha \otimes B(H_\pi)$ is nuclear and therefore the two sided ideal $A_2^\alpha(\pi)\overline{A_2^\alpha(\pi)^*}$ is nuclear. Since α is saturated, it follows that $A \otimes B(H_\pi)^{\alpha \otimes ad\pi}$ is strongly Morita equivalent (hence stably isomorphic by [2]) with the nuclear C^*-algebra $\overline{A_2^\alpha(\pi)A_2^\alpha(\pi)^*}$. Therefore, $A \otimes B(H_\pi)^{\alpha \otimes ad\pi}$ are nuclear, $\pi \in \hat{G}$. By Proposition 1, A is nuclear.

In order to state the next result let us recall a definition. A C^*-dynamical system (A, H, γ) is called minimal if there are no nontrivial hereditary C^*-subalgebras of A which are globally γ-invariant (A C^*-subalgebra $B \subset A$ is called hereditary if there is a closed left ideal $L \subset A$ such that $B = L \cap L^*$).

Proposition 3. *Let (A, G, α) be a C^*-dynamical system with G compact, separable and A separable. Assume there is a minimal C^*-dynamical system (A, H, γ) such that $\gamma_h \alpha_g = \alpha_g \gamma_h$ for every $h \in H, g \in G$. Then, if A^α is nuclear, it follows that A is nuclear.*

Proof: We shall prove that the hypotheses of the Proposition imply that the action α is saturated and then apply the preceding results. Let $\pi \in \hat{G}$. By [7, Remark 3.2] there exists an approximate identity $\{e_\lambda\}_{\lambda \in \Lambda}$ of $A_2^\alpha(\pi)^* A_2^\alpha(\pi)$ of the form $d_\lambda = \sum_{i=1}^{n_\lambda} a_i^* a_i$ where $a_i \in A_2^\alpha(\pi)$. Let $e_\lambda = [e_{ij}^\lambda]$ with $e_{ij}^\lambda \in A, 1 \le i, j \le d\pi$. Then $\sup_{\lambda \in \Lambda} e_\lambda = e$, where $e = [e_{ij}]$ is the unit of the double dual $(A_2^\alpha(\pi)^* \overline{A_2^\alpha(\pi)})''$ of $A_2^\alpha(\pi)^* A_2^\alpha(\pi)$. Since γ commutes with α, it is easy to see that $\gamma_h(e) = e$ for every $h \in H$ (we have denoted $\gamma_h(e) = [\gamma_h^{**}(e_{ij})]$). Since $e = \sup_{\lambda \in \Lambda} e_\lambda$, we have, in particular, that $e_{ii}(1 \le i \le d_\pi)$ is a lower semi-continuous element in A''. By [1, Proposition 4.1], e_{ii} are scalar multiples of the identity, say $e_{ii} = \lambda_{ii} 1, 1 \le i \le d_\pi$. We shall now prove that all $e_{ij}, 1 \le i, j \le d_\pi$, are scalar multiples of the identity.

Let H be the Hilbert space of the universal representation of A and let $\xi \in H, \xi \neq 0$ be arbitrary. Note that $H \otimes H_\pi = \sum^{d_\pi} \oplus H_i, H_i = H$ for every $1 \le i \le d_\pi$. Consider the following element ξ_{ij} of $H \otimes H_\pi$:

$$\xi_{ij} = 0 \oplus \ldots \oplus \xi \oplus 0 \oplus \ldots \oplus \xi \oplus \ldots \oplus 0.$$

Taking into account that $e_{ii} = \lambda_{ii}1, \lambda_{ii} \in \mathbb{R}_+, 1 \leq i \leq d_\pi$ we have:

$$\langle e\xi_{ij}, \xi_{ij}\rangle = \langle e_{ii}\xi, \xi\rangle + \langle e_{ij}\xi, \xi\rangle + \langle e_{ij}^*\xi, \xi\rangle + \langle e_{jj}\xi, \xi\rangle$$
$$= \langle\left(e_{ij} + e_{ij}^*\right)\xi, \xi\rangle + (\lambda_{ii} + \lambda_{jj})\|\xi\|^2$$
$$= \sup_{\lambda \in \Lambda}\langle\left(e_{ij}^\lambda + e_{ij}^{\lambda*}\right)\xi, \xi\rangle + (\lambda_{ii} + \lambda_{jj})\|\xi\|^2.$$

Therefore, $e_{ij} + e_{ij}^*$ is a self-adjoint lower semi-continuous element in A''. Applying again [1, Proposition 4.1] we infer that $e_{ij} + e_{ij}^*$ is a scalar multiple of the identity. Similarly, considering

$$\tilde{\xi}_{ij} = 0 \oplus \ldots \oplus (-\xi) \oplus 0 \oplus \ldots \oplus \xi \oplus 0 \oplus \ldots \oplus 0$$

we infer that $e_{ij} = e_{ij}^*$ is a scalar multiple of the identity. Therefore, $e_{ij} = \lambda_{ij}1, \lambda_{ij} \in \mathbb{C}$. Since $e \in [A \otimes B(H_\pi)^{\alpha \otimes ad\pi}]'' \subset A'' \otimes B(H_\pi)^{\alpha'' \otimes ad\pi}$, and π is irreducible, we obtain $e = 1$. We have then proved that $A_2^\alpha(\pi)^* \overline{A_2^\alpha(\pi)} = A \otimes B(H_\pi)^{\alpha \otimes ad\pi}$ for every $\pi \in \widehat{G}$, hence the action α is saturated. Proposition 3 follows now from Proposition 2.

References

1. D. Avitzour, Noncommutative topological dynamics I, Trans. Amer. Math. Soc. **282** (1984), 109–119.

2. L. G. Brown, P. Green, and M. A. Rieffel, Stable isomorphism and strong Morita equivalence of C^*-algebras, Pacific J. Math. **71** (1977), 349–363.

3. R. Godement, A theory of spherical functions I, Trans. Amer. Math. Soc. **73** (1952), 496–556.

4. M. B. Landstad, Algebras of spehrical functions associated with covariant systems over a compact group, Math. Scand. **47** (1980), 137–149.

5. G. K. Pedersen, *C*-Algebras and Their Automorphism Groups*, Academic Press, New York, 1979.

6. C. Peligrad, Locally compact groups actions on C^*-algebras and compact subgroups, J. Func. Anal. **76** (1988), 126–139.

7. C. Peligrad, Ergodic actions commuting with compact actions and applications to crossed products (preprint).

8. C. Peligrad, A note on nuclear crossed products (preprint).

9. J. Rosenberg, Appendix to O. Bratteli's paper on crossed products of UHF algebras, Duke Math. J. **46** (1979), 25–26.

10. R. Hoegh-Krohn, M. B. Landstad, and E. Stormer, Compact ergodic actions of automorphisms, Ann. of Math. **114** (1981), 75–86.

C. Peligrad
Department of Mathematical Sciences
University of Cincinnati
Cincinnati, Ohio 45221-0025

G. L. Price

Shifts on certain operator algebras

The purpose of this paper is to give an exposition of some of the results in [8], [9], [10], [11], and some speculations arising from these results. We do not always present results in the fullest generality in which they are known. Our main motivation for this study is R. T. Powers' paper [8] on $*$-endomorphisms on operator algebras. Although Powers considered both discrete and continuous semigroups of $*$-endomorphisms, we concentrate here on the former. For an overview of both Powers' and W. B. Arveson's beautiful work on index theory for continuous semigroups of $*$-endomorphisms, one may consult Arveson's paper [1].

§1. $*$-Endomorphisms on Factors

In what follows we shall assume that M is a factor either of Type I or Type II$_1$. We say that a unital $*$-endomorphism σ of M is a shift if σ satisfies the range condition $\cap_{k \geq 0} \sigma^k(M) = \mathbb{C}I$. Note that for any $k > 0, \sigma^k(M)$ is a subfactor of M, unitally embedded in M.

It is easy to construct examples of shifts on factors. For example, one may begin by taking a pure state ρ on the algebra $M_n(\mathbb{C})$ of $n \times n$ matrices over $\mathbb{C}$. By tensoring one may form the UHF algebra $B = \otimes_{k \geq 1} B_k$ of type n^∞, where $B_k \simeq M_n(\mathbb{C})$. Let σ be the $*$-endomorphism on B given by $\sigma(A) = I \otimes A$. More precisely, if π_k is the inclusion mapping of $M_n(\mathbb{C})$ as B_k into B, then $\sigma \circ \pi_k = \pi_{k+1}$. $\mathfrak{B}(\mathfrak{h})$ is the weak operator completion of B in the GNS construction associated with the pure symmetric product state $\omega = \otimes_{k \geq 1} \rho$ on B, and one may check that σ extends in an obvious way to a shift on $\mathfrak{B}(\mathfrak{h})$.

Conversely, suppose σ is a shift on $\mathfrak{B}(\mathfrak{h})$ leaving invariant a pure normal state ω of $\mathfrak{B}(\mathfrak{h})$, i.e., $\omega \cdot \sigma = \omega$. Let $N_1 = \sigma(\mathfrak{B}(\mathfrak{h}))'$. Then N_1 is also a type I factor (possibly finite dimensional). Letting $B_k = \sigma^{k-1}(N_1)$, one sees that the B_k are mutually commuting subfactors of $\mathfrak{B}(\mathfrak{h})$, isomorphic to N_1. This leads to the following structure theorem for shifts on $\mathfrak{B}(\mathfrak{h})$, [8].

Theorem 1. *The subfactors B_k generate $\mathfrak{B}(\mathfrak{h})$, i.e., $\{B_k : k \in \mathbb{N}\}'' = \mathfrak{B}(\mathfrak{h})$. Moreover, ω decomposes as a pure symmetric product state with respect to the decomposition $\otimes_{k \geq 1} B_k$.*

Definition 1. *Let α, β be shifts on M. Then α and β are said to be conjugate if there exists $\gamma \in \mathrm{Aut}(M)$ such that $\gamma \cdot \alpha = \beta \cdot \gamma$. α and β are outer conjugate if there is a unitary element $U \in M$ such that α and $\mathrm{Ad}(U) \cdot \beta$ are conjugate.*

There is a simple criterion for determining the outer conjugacy classes of shifts on $\mathfrak{B}(\mathfrak{h})$, [8].

Theorem 2. *Suppose α, β are shifts of $\mathfrak{B}(\mathfrak{h})$, and suppose there are pure normal states $\omega_\alpha, \omega_\beta$ which are invariant under α and β. Then α and β are outer conjugate if and only if they have the same multiplicity (the dimension of the commutant, $\alpha(\mathfrak{B}(\mathfrak{h}))'$).*

As one might expect, the study of the outer conjugacy classes of shifts on the hyperfinite II_1 factor R is more complicated than for $\mathfrak{B}(\mathfrak{h})$. There are many unresolved questions: indeed, it is not even known whether there are uncountably many outer conjugacy classes of shifts of index 2 on R (see below for a definition of index). It is known, however, that the index of a shift on R is an outer conjugacy invariant.

It is not convenient to carry over the definition of index (multiplicity) in the Type I case to the Type II_1 case: the reason for this is that there exist proper subfactors of R having trivial relative commutant. A more useful approach in the Type II case is to define the index of σ to be the subfactor index $[R : \sigma(R)]$ of $\sigma(R)$ in R, [5]. This index is a conjugacy and an outer conjugacy invariant for σ, but in contrast to the "nice" Type I case above, where there are invariant pure normal states, index is not a necessary and sufficient condition for outer conjugacy.

As an introduction to the Type II situation, we shall restrict our attention to the case where σ has index 2 (various authors have extended these results to the general integer index case, cf. [2], [3], [4], [10]). In fact, we shall further restrict our attention to the case where σ is a regular shift, which we define as follows:

Definition 2. *The shift σ is a regular shift if the subgroup $N(\sigma) = \{U \in U(R) : U\sigma^k(R)U^* = \sigma^k(R) \text{ for all } k \in \mathbb{N}\}$ of the unitary group $U(R)$ generates all of R, i.e., $N(\sigma)'' = R$.*

It is straightforward to construct a class of regular shifts of index 2 on R. These are the binary shifts constructed by Powers in [8]. Let $\{u_k : k \in \mathbb{N}\}$ be a sequence of self-adjoint unitary operators, and let $S : \mathbb{Z} \to \{1, -1\}$ be a mapping, the anti-commutation sequence, which satisfies $S_j = S_{-j}$ and $S_o = 1$. We impose the commutation relations

$$u_i u_j = S_{i-j} u_j u_i.$$

$A(S)$ is the *AF* C^*-algebra generated by the words $\Gamma(Q) = u_{i_1} u_{i_2} \ldots u_{i_n}$, where Q is an ordered n-tuple $\{i_1, i_2, \ldots, i_n\}$ of positive integers, for any $n < \infty$. Observe, for example, that $\Gamma(Q)\Gamma(Q') = \pm\Gamma(Q\nabla Q')$, where $Q\nabla Q'$ is the symmetric difference of the two sets. This follows from the anti-commutation relations and the property, $u_k^2 = I$, for all k. $A(S)$ has a tracial state τ, defined uniquely by the relations $\tau(\Gamma(Q)) = 0$ for $Q \neq \emptyset$, and $\tau(\Gamma(\emptyset)) = 1$. Of course, if σ is the constant function $\sigma(j) = 1$ for all j, then $A(\sigma)$ is commutative; and in general, $A(S)$ may or may not have a non-trivial center. More precisely ([8]):

Theorem 3. *Let* $S : \mathbb{Z} \to \{1, -1\}$ *be a sequence satisfying the conditions* $S_j = S_{-j}$ *for all* $j \in \mathbb{N}$, *and* $S_o = 1$. *Let* $A(S)$ *be the corresponding algebra generated by the commutation relations associated with* S. *Then the following conditions are equivalent.*

(i) *$A(S)$ is simple.*
(ii) *The center of $A(S)$ consists of multiples of the identity.*
(iii) *$A(S)$ has a unique normalized trace.*
(iv) *For each non-empty finite set Q of positive integers, there is an integer k such that* $U_k\Gamma(Q) = -\Gamma(Q)U_k$.

In what follows we consider only those S which satisfy the four equivalent conditions above. Such a sequence is called *primary*. For each primary S, the completion of $A(S)$ with respect to the GNS representation of the trace gives the hyperfinite II_1 factor R. The following result [11], gives a characterization of the primary anti-commutation sequences.

Theorem 4. *Let* $A(S)$ *be the* C^*-*algebra associated with an anti-commutation sequence* S. *Then* $A(S)$ *is simple if and only if the sequence* $\{S(k) : k \in \mathbb{Z}\}$ *is not periodic.*

We define a *-endomorphism σ_S on $A(S)$ by setting $\sigma_S(u_k) = u_{k+1}$, for all $k \in \mathbb{N}$. It is easy to check that σ_S extends in an obvious way to a shift on R, of index 2. A shift constructed in this manner is called a *binary shift*. Since each of the words $\Gamma(Q)$ in the u_k's lies in $N(\sigma_s)$, we easily see that σ_S is a regular shift on R. (In fact [8], $N(\sigma_S)$ consists only of scalar multiples of words.)

Theorem 5. ([8]) *If S, S' are distinct primary anticommutation sequences then σ_S and $\sigma_{S'}$ are not conjugate.*

Combining the results of the last two theorems we have:

Corollary. *There are uncountably many conjugacy classes of binary shifts on R.*

The determination of the outer conjugacy classes of binary shifts is a much more difficult issue, although some progress has been made. For instance, the first k for which $\sigma^k(R)$ has a non-trivial relative commutant (the commutant index, $k \in \{\infty, 2, 3, 4, \ldots\}$) is an outer conjugacy invariant. ($\sigma(R)$ never has a non-trivial relative commutant by a result of Jones [5].) For a given integer $k \geq 2$ it is easy to construct examples of binary shifts σ for which k is the commutant index. Hence there are at least countably many outer conjugacy classes of binary shifts on R.

Some recent papers ([2], [4]) have identified finer invariants for outer conjugacy. These invariants emerge from the observation that for a regular binary shift σ on R with finite commutant index, σ also induces a binary shift on the algebra generated by the relative commutants of $\sigma^k(R), \sigma^{k+1}(R), \ldots$. The anti-commutation sequence corresponding to the latter binary shift is an outer conjugacy invariant for the original shift [2]. This result does not resolve the question of cardinality, however, since only countably many binary shifts (those for which the sequence $\{S_k : k \in \mathbb{N}\}$ is eventually periodic) have finite commutant indices.

However, we conjecture that there are uncountably many outer conjugacy classes. The evidence we have for this assertion is the observation, based on computing examples, that if α, β are outer conjugate shifts, their corresponding sequences $S(\alpha), S(\beta)$ exhibit "similar behavior". As just noted, $S(\alpha)$ is eventually periodic if and only if the same is true of $S(\beta)$. But there appear to be more subtle properties which are preserved under outer conjugacy. For example, suppose we associate with an anti-commutation sequence $S(\alpha)$ the real number

$$\gamma_\alpha = \sum_{k=1}^{\infty} \frac{s_k}{2^k}.$$

Suppose γ_α is a Liouville number. Then we suspect that γ_α must also be Liouville. We plan to consider this problem elsewhere.

Finally, the question arises as to whether every regular shift σ of index 2 on R is a binary shift. The answer is negative, even if one imposes the requirement that $N(\sigma)$ consists of elements whose square is a scalar multiple of the identity (this is automatic for binary shifts). Our construction of a counterexample amounts to gluing infinitely many copies of binary shifts [11]. Choose self-adjoint unitaries $u_{ij}, i, j \in \mathbb{N}$, which we shall assume pairwise either to commute or anti-commute. We define a $*$-endomorphism σ on the algebra generated by the u_{ij}'s by $\sigma(u_{ij}) = u_{ij+1}$. We also assume that for $i, j \in \mathbb{N}$,

$$u_{ij} = u_{i+1,j}\sigma(u_{i+1,j}). \qquad (*)$$

Define a trace on the algebra of polynomials in the u_{ij}'s by $\tau(w) = 0$ if w is a non-trivial word.

Theorem 6. *([11]) There exists an assignment of anti-commutation relations, compatible with $(*)$, for the elements u_{ij}, such that R is the von Neumann algebra completion $\{u_{ij} : i, j \in \mathbb{N}\}''$ with respect to the trace τ. The $*$-endomorphism σ extends to a shift of index 2 on R, but σ is not a binary shift on R.*

§2. Endomorphisms on Certain Operator Algebras

As noted above, the shifts on $\mathfrak{B}(\mathfrak{h})$ are well-understood, so it would seem natural to progress to the study of $*$-endomorphisms of $\mathfrak{U} = C(X, \mathfrak{B}(\mathfrak{h}))$, the C^*-algebra of uniformly continuous functions from a compact Hausdorff space X to $\mathfrak{B}(\mathfrak{h})$. (In what follows we shall assume that X is separable.) Another reason for studying these algebras is that the structure of the group of $C(X)$-automorphisms of $\mathfrak{U}$ is well-understood from the beautiful work of Kadison-Ringrose [6], and Lance [7]. One might hope for analogous results for $*$-endomorphisms.

By a $*$-endomorphism σ of $\mathfrak{U}$ we mean a $*$-homomorphism from the algebra into itself, preserving the identity ($\sigma(\underline{I}) = \underline{I}$, where $\underline{I}(x) = I$, for all $x \in X$). By a $C(X)$-endomorphism on $\mathfrak{U}$ we mean a $*$-endomorphism σ which satisfies $\sigma(sA) = s\sigma(A)$, for all functions $s \in \mathfrak{U}$ taking values in the scalar operators. Of course, this includes the class of $C(X)$-automorphisms mentioned above. Finally, a $C(X)$-shift is a $C(X)$-endomorphism satisfying the range condition

$$\bigcap_{k \geq 0} \sigma^k(\mathfrak{U}) = C(X).$$

(Observe that $\sigma^k(s\underline{I}) = s\underline{I}$ for all k, so that one always has $C(X) \subset \sigma^k(\mathfrak{U})$). Arguing as in [6], one may show that a $C(X)$-endomorphism is induced locally in the sense that

103

there are $*$-endomorphisms $\sigma_x, x \in X$, of $\mathfrak{B}(\mathfrak{h})$ such that

$$\sigma(A)(x) = \sigma_x(A(x)).$$

By evaluating σ on constant functions in $\mathfrak{U}$ one may show easily that the mapping $x \to \sigma_x$ is continuous in the pointwise norm topology, and that the index of σ_x (as defined in the previous section) is constant on connected components of X.

If σ_x has index 1 then it is an automorphism, and $\sigma_x = Ad(U)$, for some unitary operator U. If σ_x has (finite) higher index n^2, there are isometries $V_1, \ldots, V_n$ satisfying

$$V_i^* V_j = \delta_{ij} I,$$

$$\sum_{i=1}^{n} V_i V_i^* = I,$$

and

$$\sigma_x(A) = \sum_{i=1}^{n} V_i A V_i^*.$$

In other words, there is an action of the Cuntz algebra, 0_n, on $\mathfrak{B}(\mathfrak{h})$. If σ_x has infinite index there is a similar action of 0_∞ on $\mathfrak{B}(\mathfrak{h})$ which implements the $*$-endomorphism. Alternatively, if $x \in X$ is fixed and $y \in X$ is sufficiently close to X, there is a unitary element $U_y \in \mathfrak{B}(\mathfrak{h})$ such that $\sigma_y = Ad(U_y) \cdot \sigma_x$. The following theorem [9] is reminiscent of a result in [7] on the structure of $C(X)$-automorphisms.

Theorem 7. *Let $\sigma = \oplus_{x \in X} \sigma_x$ be a $C(X)$-endomorphism of $\mathfrak{U} = C(X, \mathfrak{B}(\mathfrak{h}))$ on a connected space X, for which each σ_x is a $*$-endomorphism of finite index on $\mathfrak{B}(\mathfrak{h})$. Suppose there is a pure normal state ρ on $\mathfrak{B}(\mathfrak{h})$ such that $\rho = \rho \cdot \sigma_x$, for all $x \in X$. Then for each $x \in X$ there is a neighborhood N of x such that*

(i) *$\sigma_y = Ad(U_y) \cdot \sigma_x$, for $y \in N$, and*

(ii) *the U_y vary continuously in norm for $y \in N$.*

In [7] E. C. Lance considers the quotient of the group of $C(X)$-automorphisms by the $C(X)$-inner automorphisms. He shows that $\mathrm{Aut}_{C(X)}(\mathfrak{U})/\mathrm{Inn}_{C(X)}(\mathfrak{U})$ is isomorphic to $H^2(X, \mathbb{Z})$. Now let us suppose that X is connected, so that the local $*$-endomorphisms σ_x of a $C(X)$-endomorphism all have the same index. We shall say that σ is *inner* if there is a $*$-endomorphism α of $\mathfrak{B}(\mathfrak{h})$ and a unitary operator U in $\mathfrak{U}$ such that $\sigma = Ad(U) \cdot \alpha$, i.e., for $x \in X, \sigma_x = Ad(U(x)) \cdot \alpha$.

If σ is any $C(X)$-endomorphism, the theorem above ensures that there is an open cover $\{N_i\}_{i \in \mathfrak{J}}$ of X, unitary-valued functions $U_i \in C(N_i, \mathfrak{B}(\mathfrak{h}))$, and a $*$-endomorphism

α of $\mathfrak{B}(\mathfrak{h})$ such that $\sigma_x = \mathrm{Ad}(U_i(x)) \cdot \alpha$, for $x \in N_i$. For simplicity assume that $M = \alpha(\mathfrak{B}(\mathfrak{h}))'$ is finite-dimensional. If $\Lambda_{ij} = U_i^* U_j$ on $N_{ij} = N_i \cap N_j$, easy observations show that $\Lambda_{ij} \in C(N_{kj}, M)$ and that the (non-commutative) cocycle condition $\Lambda_{ij}\Lambda_{jk} = \Lambda_{ik}$ on $N_{ijk} = N_i \cap N_j \cap N_k$ is satisfied. Again for simplicity replace Λ_{ij} with $\lambda_{ij} = \det(\Lambda_{ij})$. Then $\{\lambda_{ij}\}$ is a π^1-valued cocycle on X. If σ is inner, the corresponding cocycle $\{\lambda_{ij}\}$ is a coboundary. Following the line of argument in [7], one can show that $H^2(X, \mathbb{Z})$ is an obstruction to a $*$-endomorphism being inner. The full obstruction, which we will not consider here, is a problem in non-commutative set cohomology.

Finally, there is a very basic question about $C(X)$-endomorphisms which seems very difficult to resolve, namely, when is a $C(X)$-endomorphism a $C(X)$-shift? It is straightforward to show that σ is a shift if each of its components σ_x is a shift. It is perhaps surprising that the converse is false [9]. Indeed, we have the following example. Let α be the shift on $B = \otimes_{k \geq 1} B_k$ described in the previous section. Let $C = B_o \otimes B$, where B_o is another copy of $M_n(\mathbb{C})$, and let $\alpha_o = \iota \otimes \alpha$. Clearly, α_o is not a shift on $\mathfrak{B}(\mathfrak{h})$. Now, let $T \in B_o \otimes B_1$ be a self-adjoint unitary with the property that $\mathrm{Ad}(e^{itT}) \cdot \alpha_o$. Setting $X = [0, \pi/2]$ and $\sigma = \otimes_{t \in X} \alpha_t$, σ is a shift on $C(X, \mathfrak{B}(\mathfrak{h}))$.

It would be very interesting to obtain necessary and sufficient conditions for a $C(X)$-endomorphism to be a $C(X)$-shift.

References

1. W. B. Arveson, An addition formula for the index of semigroups of endomorphisms of $\mathfrak{B}(\mathfrak{h})$, Pac. J. Math. **137** (1989), 19–36.

2. D. Bures and H. S. Yin, Shifts on the hyperfinite factor of type II$_1$, J. Operator Theory **20** (1988), 91–106.

3. M. Choda, Shifts on the hyperfinite II$_1$-factor, J. Operator Theory **17** (1987), 223-235.

4. M. Enomoto and Y. Watatani, Powers' binary shifts on the hyperfinite factor of type II$_1$, Proc. Amer. Math. Soc., to appear.

5. V. F. R. Jones, Index for subfactors, Invent. Math. **72** (1983), 1–25.

6. R. V. Kadison and J. R. Ringrose, Derivations and automorphisms of operator algebras, Comm. Math. Phys. 4 (1967), 32–63.

7. E. C. Lance, Automorphisms of certain operator algebras, Amer. J. Math. **91** (1969), 160–174.

8. R. T. Powers, An index theory for semigroups of $\mathfrak{B}(\mathfrak{h})$ and type II$_1$ factors, Can. J. Math. **40** (1988), 86–114.

9. G. L. Price, Endomorphisms of certain operator algebras, Publ. R.I.M.S., Kyoto University **25** (1989), 45–57.

10. G. L. Price, Shifts of integer index on the hyperfinite II$_1$ factor, Pacific J. Math. **132** (1988), 379–390.

11. G. L. Price, Shifts on type II$_1$ factors, Can. J. Math. **39** (1987), 492–511.

Geoffrey L. Price

Department of Mathematics

U. S. Naval Academy

Annapolis, Maryland 21402

Supported in part by grants from the National Science Foundation and the Naval Academy Research Council.

S. Teleman

Measure–theoretic properties of the central topology on the set of the factorial states of a C*-algebra. Central reduction

On the set of the extreme points of a compact convex set various topologies have been introduced having relevance to the boundary measures (see [1], [4], [6], [8], [9], [13], [14], [15], [16], [21], [28], [29], [30], [31], [33], [35]). As a result, a general topological measure theory has been obtained, which contains, as particular cases, both the Borel measure theory in Polish spaces, as well as the Radon measure theory in arbitrary compact spaces; and also, this allowed the development of a "non-commutative Borel analysis" over arbitrary C^*-algebras (see [36]).

In the theory of C^*-algebras an important role is also played by the set of the factorial (primary) states of a given C^*-algebra.

In this paper we shall introduce a natural topology in the set of $F(A)$ of the factorial states of an arbitrary C^*-algebra A, and we shall show that the central measures on the set $E_0(A)$ of the quasi-states of A, which represent states of A, induce regular Borel measures on $F(A)$ with whose help a spatial central reduction theory for the representations of A can be developed.

We refer to [10], [11], [12], [22], [24], and [27] for general facts concerning the theory of C^*-algebras. In ([22], Ch. IV) one can find a deep analysis of Reduction Theory for the separable case.

In the present paper no separability conditions are assumed. Theorem 33 is the main result of the paper.

§1. Introduction

We shall denote by $E_0(A)$ the convex set of all the *quasi-states* of A, endowed with the topology $\sigma(A^*; A)$, for which it is compact; $E(A) = \{f \in E_0(A); \|f\| = 1\}$ is the set of the *states* of A, which is a convex G_δ-subset of $E_0(A)$. The set $E(A)$ is compact if, and only if, A possesses the unit element, which we shall denote by 1.

The set of the extreme points of $E_0(A)$ is given by the equality

$$ex\, E_0(A) = P(A) \cup \{0\},$$

where $P(A)$ is the set of the pure states of A (see [12], Proposition 2.5.5).

For any positive $f \in A^*_+ = \{f \in A^*; f \geq 0\}$, we shall denote by $\pi_f/A \to \mathcal{L}(H_f)$ the representation of A on the associated Hilbert space H_f, by $\theta_f : A \to H_f$ the associated linear mapping, and by ξ_f^0 the associated cyclic vector, according to the GNS-construction. We have

$$f(a) = \left(\pi_f(a)\xi_f^0|\xi_f^0\right), \theta_f(a) = \pi_f(a)\xi_f^0, \quad a \in A,$$
$$\|f\| = \|\xi_f^0\|^2, \quad f \in A^*_+.$$

The state $f \in E(A)$ is said to be *factorial* (or *primary*) if $\pi_f(A)''$ is a factor. We denote by $F(A)$ the set of all factorial states of A. We obviously have that

$$F(A) \supset P(A).$$

We refer to [3], [7] for recent results relating topological properties of $F(A)$ to structural properties of A.

In order to develop a working topological theory for the boundary measures induced on $P(A)$ by the maximal (orthogonal) measures on $E_0(A)$ one has first to find a suitable topology on $P(A)$. In this respect, we refer to [5], [6], [28], [29], [30], [31], [32], [33], [36], and [37] for the corresponding definitions, results, and applications obtained. In particular, an irreducible spatial disintegration theory has been obtained for the (cyclic) representations of A, and for their extensions to the Baire and the Borel enveloping C^*-algebras of A. We point out here only the fact that several topologies have been introduced on $P(A)$, compatible with the maximal (orthogonal) measures, thus allowing several regular Borel extensions of the induced boundary measures, the maximal orthogonal topology being, so far, the strongest in this family (see [6], [33]).

We shall define below the *central topology* on $F(A)$ and we shall show that any central Radon probability measure on $E_0(A)$, whose barycenter is in $E(A)$, induces a regular Borel probability measure on $F(A)$.

For the central topology, the space $F(A)$ satisfies the (T_1) separation axiom; and it is quasi-compact if, and only if, A possesses the unit element.

For any topological space X we shall denote by $\mathcal{B}_0(X)$ (or by $\mathcal{B}_0(X;\tau)$ when the topology τ of X is to be emphasized) the σ-algebra of the *Baire* subsets of X: it is the smallest σ-algebra of subsets of X, containing all closed G_δ-subsets of X; by $\mathcal{B}(X)$ (or by $\mathcal{B}(X;\tau)$, when the topology τ of X is to be emphasized) we shall denote the σ-algebra of the *Borel* subsets of X; it is the smallest σ-algebra of subsets of X, containing all closed subsets of X. Of course, we have the inclusion

$$\mathcal{B}_0(X) \subset \mathcal{B}(X).$$

If X is metrizable, then

$$\mathcal{B}_0(X) = \mathcal{B}(X),$$

but for most compact spaces X we have $\mathcal{B}_0(X) \neq \mathcal{B}(X)$.

By $C(X)$ we shall denote the algebra of all continuous complex functions on X, whereas $C^b(X)$ $(\subset C(X))$ will stand for the C^*-algebra of all bounded continuous complex functions on X, endowed with the supnorm.

For any compact (Hausdorff) topological space K we shall denote by $\mathcal{M}_+(K)$ the space of all positive Radon measures on K, endowed with the vague topology; i.e., the topology induced by $\sigma(C(K)^*; C(K))$, since $\mathcal{M}_+(K)$ can be identified with a subset of the dual Banach space $C(K)^*$ of $C(K)$. By $\mathcal{M}_+^1(K)$ we shall denote the compact convex subset of $\mathcal{M}_+(K)$, consisting of all Radon probability measures on K.

If K is a convex compact subset of a Hausdorff locally convex topological real vector space X, then for any $\mu \in \mathcal{M}_+(K)$ we can define its *resultant* $r(\mu) \in X$ by the formula

$$x^*(r(\mu)) = \int_K x^*(x) \, d\mu(x), \qquad x^* \in X^*;$$

it always exists and it is unique. If $\mu \in \mathcal{M}_+^1(K)$, then $r(\mu) \in K$; in this case it is called the *barycenter* of μ and it is denoted by $b(\mu)$.

If we denote by $A(K)$ the real Banach space of all continuous affine real functions on K, then, by taking into account the fact that the space of all the restrictions to K of the functions $x^* \in X^*$ is uniformly dense in $A(K)$, we immediately infer that we have the formula

$$h(b(\mu)) = \int_K h(x) \, d\mu(x), \qquad h \in A(K).$$

We shall also denote $\mathcal{M}_+^1(K; x) = \{\mu \in \mathcal{M}_+^1(K); b(\mu) = x\}$. By $S(K)$ we shall denote the sup-cone of all convex continuous real functions on the compact convex set K.

§2. Subcentral Measures

I. For any $f_0 \in E_0(A)$, and any Radon measure μ on $E_0(A)$, whose resultant $r(\mu) = f_0$, one has the *Tomita mapping* $K_\mu : L^\infty(\mu) \to \mathcal{L}(H_{f_0})$, which is linear, positive, $(\sigma(L^\infty(\mu); L^1(\mu)),$ weak operator)-continuous and $\mathrm{im}\, K_\mu \subset \pi_{f_0}(A)'$; it is defined by the equality

$$\int_{E_0(A)} \phi \lambda_A(a)\, d\mu = (K_\mu([\phi])\pi_{f_0}(a)\xi^0_{f_0} | \xi^0_{f_0}), \quad a \in A,$$

where $[\phi]$ is the class in $L^\infty(\mu)$ of the bounded Borel measurable function $\phi : E_0(A) \to \mathbb{C}$, and $\lambda_A(a) : E_0(A) \to \mathbb{C}$ is given by $\lambda_A(a)(f) = f(a), f \in E_0(A)$.

We refer to [28] for the basic properties of K_μ, which we shall use below.

The measure μ is said to be *orthogonal* if for any $M \in \mathcal{B}(E_0(A))$ we have that

$$r(\chi_M \mu) \perp r(\chi_{CM} \mu);$$

i.e., for any $f \in A^*_+$, such that

$$f \le r(\chi_M \mu) \text{ and } f \le r(\chi_{CM} \mu),$$

we have $f = 0$.

The Radon measure μ is orthogonal if, and only if, the mapping K_μ is multiplicative. In this case $C_\mu \overset{\mathrm{def}}{=} \mathrm{im}\, K_\mu$ is an abelian von Neumann sub-algebra of $\pi_{f_0}(A)'$. We denote by e_μ the projection onto $\overline{C_\mu \xi^0_{b(\mu)}}$.

Conversely, for any state $f_0 \in E(A)$ and any abelian von Neumann sub-algebra C of $\pi_{f_0}(A)'$ there exists a unique orthogonal Radon probability measure $\mu \in \mathcal{M}^1_+(K)$, such that $b(\mu) = f_0$ and $C_\mu = C$.

We shall denote by $\Omega(E_0(A); f_0)$ the set of all orthogonal Radon probability measures μ on $E_0(A)$, such that $b(\mu) = f_0$; also, it will be convenient to denote by $\Omega(E_0(A))$ the set $\cup\{\Omega(E_0(A); f_0); f_0 \in E_0(A)\}$ of all orthogonal Radon probability measures μ on $E_0(A)$.

If A possesses the unit element ($1 \in A$), then $\Omega(E(A); f_0)$ will stand for the set of all orthogonal Radon probability measures μ on $E(A)$, such that $b(\mu) = f_0 \in E(A)$, whereas $\Omega(E(A))$ will denote the set $\cup\{\Omega(E(A); f_0); f_0 \in E(A)\}$ of all orthogonal Radon probability measures on $E(A)$.

In particular, to any state $f_0 \in E(A)$ we can associate its corresponding *central measure* $\mu_{f_0} \in \mathcal{M}^1_+(E_0(A))$, which is the orthogonal Radon probability measure corresponding to the center $\pi_{f_0}(A)'' \cap \pi_{f_0}(A)'$ of $\pi_{f_0}(A)'$, and whose barycenter is at f_0.

An orthogonal measure $\mu \in \mathcal{M}_+(E_0(A))$ is said to be *subcentral* if $C_\mu \subset \pi_{f_0}(A)'' \cap \pi_{f_0}(A)'$, where $f_0 = r(\mu)$.

110

Lemma 1. *For any measure $\mu \in \mathcal{M}_+(E_0(A))$ the following are equivalent*

(a) *μ is subcentral;*

(b) *for any $M \in \mathcal{B}(E_0(A))$ the representations $\pi_{r(\chi_M \mu)}$ and $\pi_{r(\chi_{CM}\mu)}$ are disjoint.*

Proof: (a) $\Rightarrow$ (b). We have $f_0 \overset{\text{def}}{=} r(\mu) \in A_+^*$ and

$$(K_\mu([\phi])\pi_{f_0}(a)\xi_{f_0}^0 | \xi_{f_0}^0) = \int_{E_0(A)} \phi \lambda_A(a) \, d\mu, \phi \in \mathcal{L}^\infty \left(E_0(A), \mathcal{B}(E_0(A)) \right).$$

Therefore, we have

$$r(\chi_M \mu)(a) = \left(K_\mu \left([\chi_M] \right) \pi_{f_0}(a)\xi_{f_0}^0 | \xi_{f_0}^0 \right), \quad a \in A,$$
$$r(\chi_{CM}\mu)(a) = \left(K_\mu \left(1 - [\chi_M] \right) \pi_{f_0}(a)\xi_{f_0}^0 | \xi_{f_0}^0 \right), \quad a \in A,$$

for any $M \in \mathcal{B}(E_0(A))$. By hypothesis, for $e \overset{\text{def}}{=} K_\mu([\chi_M])$, we have $e \in \pi_{f_0}(A)'' \cap \pi_{f_0}(A)'$, and the representations $\pi_{r(\chi_M \mu)}$, resp., $\pi_{r(\chi_{CM}\mu)}$, are unitarily equivalent to the representations $\pi_1 : A \ni a \mapsto \pi_{f_0}(a)e$, in eH_{f_0}, resp., $\pi_2 : A \ni a \mapsto \pi_{f_0}(a)(1-e)$, in $(1-e)H_{f_0}$. Assume that there exists a partial isometry $v : eH_{f_0} \to (1-e)H_{f_0}$, such that $v^*v \leq e, vv^* \leq 1 - e$ and

$$v\pi_{f_0}(a) = \pi_{f_0}(a)v, \quad a \in A.$$

Then we have $ve = v, (1-e)v = v$ and, by virtue of Kaplansky's Theorem, there exists a net $a_\alpha \in A$, such that $\pi_{f_0}(a_\alpha) \to e$. We infer that

$$v = ve = \lim_\alpha v\pi_{f_0}(a_\alpha) = \lim_\alpha \pi_{f_0}(a_\alpha)v = ev = 0,$$

and this shows that $\pi_{r(\chi_M \mu)}$ and $\pi_{r(\chi_{CM}\mu)}$ are disjoint.

(b) $\Rightarrow$ (a). First of all, from (b) one immediately infers that μ is orthogonal. If μ were not subcentral, then one would have that $\mathcal{C}_\mu \subset \pi_{f_0}(A)''$ and then one could find a set $M \in \mathcal{B}(E_0(A))$, such that $e \overset{\text{def}}{=} K_\mu([\chi_M]) \notin \pi_{f_0}(A)''$. Of course, e is a projection in $\pi_{f_0}(A)'$, and we have that

$$e\pi_{f_0}(A)'(1-e) \neq \{0\}$$

(otherwise, e would be in $\pi_{f_0}(A)''$). From the Comparison Theorem we infer that there exists a central projection $g \in \pi_{f_0}(A)'' \cap \pi_{f_0}(A)'$, such that $ge \prec g(1-e)$ and $(1-g)e \succ (1-g)(1-e)$. Then either $ge \neq 0$, or $(1-g) \cdot (1-e) \neq 0$. In the first case, there exists a partial isometry $u \neq 0$, such that $u^*u = ge, uu^* \leq g(1-e)$, and the mapping

$$\pi_{f_0}(a)ge \mapsto u\pi_{f_0}(a)u^*, \quad a \in A,$$

111

is a non-zero correctly defined unitary equivalence of a subrepresentation of π_1 with a subrepresentation of π_2. These are, therefore, not disjoint. The second case can be treated similarly.

Remark 1. The fact that two representations π_1, π_2 of A are disjoint is usually denoted by $\pi_1 \, \overset{\shortmid}{} \, \pi_2$. If $f_1, f_2 \in A_+^*$, then they are said to be *disjoint*, and one writes $f_1 \, \overset{\shortmid}{} \, f_2$, if $\pi_{f_1} \, \overset{\shortmid}{} \, \pi_{f_2}$ (see [12], 5.2.2.; [22], p. 65).

Remark 2. The preceding Lemma extends Proposition 20 from [25] to the slightly more general case of a C^*-algebra which is not assumed to possess a unit element. See, also, ([10], Proposition 4.2.9).

II. The following theorem extends to the general, possibly non-separable case, a theorem belonging to W. Wils (see [40], [41]; and also [25] Theorem 27). Also, we do not assume the existence of the unit element.

Theorem 1. *For any state $f_o \in E(A)$, the corresponding central measure μ_{f_o} is the greatest (with respect to the Choquet-Meyer order relation) Radon probability measure on $E_0(A)$, whose barycenter is at f_0, and which is dominated (with respect to the Choquet-Meyer order relation) by any Choquet maximal Radon probability measure, whose barycenter is at f_0.*

For the proof we need some lemmas, which are interesting in themselves. In presenting them, we follow C. F. Skau, except that we do not assume A to possess the unit element (see [25], Lemmas 22, 23, 24, and 25). Lemma 2 below is a slight improvement of a Theorem of G. Choquet (see [20], Ch. XI, §1.8; [10], Proposition 4.1.1.; [28], Proposition 1.15).

Lemma 2. *Let K be any compact convex set in a Hausdorff locally convex topological real vector space. Then for any $x_0 \in K$ and any $\mu \in \mathcal{M}_+^1(K; x_0)$ there exists an increasing net $(\nu_{\mathcal{L}})$ in the set $\mathcal{F}_+^1(K; x_0)$ of all Radon probability measures on K, having finite supports, and whose barycenters are at x_0, such that*

(a) $\sup\{\nu_{\mathcal{L}}(f); \mathcal{L}\} = \mu(f), f \in S(K)$;
(b) $\lim_{\mathcal{L}} \nu_{\mathcal{L}} = \mu$ in $\sigma(C(K)^*; C(K))$.

Proof: (i) Let $f \in C(K)$ and $\epsilon > 0$ be given. Then there exists a cover $K_i, 1 \leq i \leq n$, of K, by compact convex subsets of K, such that

$$\overset{n}{\underset{i=1}{\cup}} K_i = K \text{ and } x', x'' \in K_i \Rightarrow |f(x') - f(x'')| < \epsilon.$$

Let $L_1 = K_1$ and $L_i = K_i \setminus (L_1 \cup \ldots \cup L_{i-1}), 2 \leq i \leq n$. Let $J = \{i; 1 \leq i \leq n, \mu(L_i) > 0\}$ and define $\nu = \sum_{i \in J} \mu(L_i) \epsilon_{x_i}$, where $x_i = b(\mu(L_i)^{-1} \chi_{L_i} \mu), i \in J$. We obviously have $x_i \in K_i, i \in J$. It follows that

$$\left| \int_K f \, d\nu - \int_K f \, d\mu \right| = \left| \sum_{i \in J} \mu(L_i) f(x_i) - \sum_{i \in J} \int_{L_i} f \, d\mu \right|$$

$$\leq \sum_{i \in J} \int_{L_i} |f(x_i) - f(x)| \, d\mu(x) < \epsilon \sum_{i=1}^n \mu(L_i) = \epsilon.$$

(ii)   Let us now consider any cover $\mathcal{L} = \{L_1, L_2, \ldots, L_m\}$ of K, by Borel subsets $L_i \subset K$, such that $L_i \cap L_j = \emptyset$, for $i \neq j, i, j = 1, 2, \ldots, n$. For $i \in J = \{i; 1 \leq i \leq m, \mu(L_i) > 0\}$, let us define $x_i = b(\mu(L_i)^{-1} \chi_{L_i} \mu)$ and $\nu_{\mathcal{L}} = \sum_{i \in J} \mu(L_i) \epsilon_{x_i}$. For any convex continuous real function f on K we shall have

$$\nu_{\mathcal{L}}(f) = \sum_{i \in J} \mu(L_i) f(x_i) \leq \sum_{i \in J} \mu(L_i) \mu(L_i)^{-1} \int_{L_i} f \, d\mu = \int_K f \, d\mu,$$

and this shows that $\nu_{\mathcal{L}} \prec \mu$. Of course, this implies that $\nu_{\mathcal{L}} \in \mathcal{M}_+^1(K; x_0)$.

(iii)   Let $\mathcal{L}', \mathcal{L}''$ be two such covers of K, as above:

$$\mathcal{L}' = \{L_1', L_2', \ldots, L_m'\}, \mathcal{L}'' = \{L_1'', L_2'', \ldots, L_n''\}.$$

We shall now consider the cover

$$\mathcal{L} = \{L_i' \cap L_j''; 1 \leq i \leq m, 1 \leq j \leq n\}.$$

Let $J = \{(i, j); \mu(L_i' \cap L_j'') > 0\}$ and define

$$\mu_{ij} = \mu(L_i' \cap L_j'')^{-1} \chi_{L_i \cap L_j} \mu, x_{ij} = b(\mu_{ij}),$$

for any $(i, j) \in J$. We shall then have

$$\nu_{\mathcal{L}} = \sum_{(i,j) \in J} \mu(L_i' \cap L_j'') \epsilon_{x_{ij}}.$$

We shall prove that $\nu_{\mathcal{L}'} \prec \nu_{\mathcal{L}}$ and $\nu_{\mathcal{L}''} \prec \nu_{\mathcal{L}}$. Indeed, for $J' = \{i; 1 \leq i \leq m, \mu(L_i') > 0\}$, let us define

$$\nu_i = \mu(L_i')^{-1} \sum_j \mu(L_i' \cap L_j'') \epsilon_{x_{ij}} \quad i \in J',$$

113

where the sum extends over all $j, 1 \le j \le n$, such that $\mu(L_i' \cap L_j'') > 0$. We have $b(\nu_i) = b(\mu(L_i')^{-1}\chi_{L_i}\mu), i \in J'$. Indeed, we have

$$b(\nu_i) = \mu(L_i')^{-1} \sum_j \mu(L_i' \cap L_j'')x_{ij}$$

and, therefore, for any $h \in A(K)$, we have

$$h(b(\nu_i)) = \mu(L_i')^{-1} \sum_j \mu(L_i' \cap L_j'')h(x_{ij})$$

$$= \mu(L_i')^{-1} \sum_j \mu(L_i' \cap L_j'')\mu(L_i' \cap L_j'')^{-1} \int_{L_i' \cap L_j''} h \; d\mu$$

$$= \mu(L_i')^{-1} \int_{L_i'} h \; d\mu = \left(\mu\left(L_i'\right)^{-1}\chi_{L_i'}\mu\right)(h) = h\left(b\left(\mu(L_i')^{-1}\chi_{L_i'}\mu\right)\right),$$

and this shows that

$$b(\nu_i) = b\left(\mu(L_i')^{-1}\chi_{L_i'}\mu\right), \quad i \in J'.$$

For any $f \in S(K)$ we shall have ($x_k' \stackrel{\text{def}}{=} b(\nu_i), i \in J'$):

$$\nu_{\mathcal{L}'}(f) = \sum_{i \in J'} \mu(L_i')f(x_i')$$

$$= \sum_{i \in J'} \mu(L_i')f(b(\nu_i)) \le \sum_{i \in J'} \mu(L_i')\mu(L_i')^{-1} \sum_j \mu(L_i' \cap L_j'')f(x_{ij})$$

$$= \sum_{(i,j) \in J} \mu(L_i' \cap L_j'')f(x_{ij}) = \nu_{\mathcal{L}}(f),$$

and this shows that $\nu_{\mathcal{L}'} \prec \nu_{\mathcal{L}}$. Similarly, we obtain that $\nu_{\mathcal{L}''} \prec \nu_{\mathcal{L}}$.

(iv) From (ii) and (iii) we infer that $(\nu_{\mathcal{L}})$ is an increasing net in $\mathcal{F}_+^1(K; x_0)$, such that $\nu_{\mathcal{L}}(f) \le \mu(f)$, for any $\mathcal{L}$ and any $f \in S(K)$. From (i) we infer that

$$\sup\{\nu_{\mathcal{L}}(f); \mathcal{L}\} = \mu(f), \qquad f \in S(K).$$

It follows that we have $\lim_{\mathcal{L}} \nu_{\mathcal{L}}(f) = \mu(f)$, for any $f \in S(K)$. Since the vector subspace $S(K) - S(K) \subset C(K; \mathbf{R})$ is uniformly dense in $C(K; \mathbf{R})$, we infer that

$$\lim_{\mathcal{L}} \nu_{\mathcal{L}}(f) = \mu(f),$$

for any $f \in C(K; \mathbf{R})$ and, therefore, $\lim_{\mathcal{L}} \nu_{\mathcal{L}} = \mu$ in the topology $\sigma(C(\dot{K})^*; C(K))$. The Lemma is proved.

Remark 1. Measures having finite supports are also called *simple*.

Remark 2. Part (a) of the Lemma shows that $\mu = \sup\{\nu_{\mathcal{L}}; \mathcal{L}\}$ in the set $\mathcal{M}_+^1(K; x_0)$, partially ordered by the Choquet-Meyer order relation.

The following Lemma is an adaptation of Lemma 2 to the case of the orthogonal measures. It is a slight improvement of Lemma 23 from [25].

Lemma 3. *Let μ be an orthgonal Radon probability measure on $E_0(A)$, whose barycenter is a state $(b(\mu) \in E(A))$. Then there exists an increasing net $(\mu_{\mathcal{C}})$ in the set $\Omega(E_0(A); b(\mu))$, consisting of orthogonal Radon probability measures on $E_0(A)$, and having finite supports, such that*

(a) $\sup\{\mu_{\mathcal{C}}(f); \mathcal{C}\} = \mu(f), f \in S(E_0(A))$;

(b) $\lim_{\mathcal{C}} \mu_{\mathcal{C}} = \mu$ *in* $\sigma(C(E_0(A))^*; C(E_0(A)))$.

(As above, $\Omega(E_0(A), b(\mu))$ is equipped with the Choquet–Meyer order relation; or, equivalently, with the Bishop–de Leeuw order relation (see [28], Theorem 3.5).)

Proof: Let $\mathcal{C}_\mu$ be the abelian von Neumann subalgebra of $\pi_{b(\mu)}(A)'$, corresponding to μ, and denote by $e_\mu \in \mathcal{C}'_\mu$ the projection onto the subspace $\overline{\mathcal{C}_\mu \xi^0_{b(\mu)}} \subset H_{b(\mu)}$. Let $\mathcal{J}$ be the set of all finite dimensional von Neumann subalgebras of $\mathcal{C}_\mu$, partially ordered by inclusion. For any $\mathcal{C} \in \mathcal{J}$, let us denote by $e_{\mathcal{C}} \in \mathcal{C}'$ the projection onto $\overline{\mathcal{C} \xi^0_{b(\mu)}}$ and let $\mu_{\mathcal{C}}$ be the unique orthogonal Radon probability measure on $E_0(A)$, such that $b(\mu_{\mathcal{C}}) = b(\mu)$ and $im\ K_{\mu_{\mathcal{C}}} = \mathcal{C}$ (see [28], Theorem 3.2). It is obvious that $\mathcal{J}$ is an increasing net and that the net $\mathcal{J} \ni \mathcal{C} \to e_{\mathcal{C}}$ converges, in the strong operator topoogy, to e_μ.

On the other hand, from $\mathcal{C} \subset \mathcal{C}_\mu, \mathcal{C} \in \mathcal{J}$, we infer that $\mu_{\mathcal{C}} \ll \mu$; whereas, from $\mathcal{C}_1 \subset \mathcal{C}_2, \mathcal{C}_1, \mathcal{C}_2 \in \mathcal{J}$, we infer that $\mu_{\mathcal{C}_1} \ll \mu_{\mathcal{C}_2}$ (or, equivalently, $\mu_{\mathcal{C}} \prec \mu$; resp., $\mu_{\mathcal{C}_1} \prec \mu_{\mathcal{C}_2}$; see [28], Theorem 3.5).

From the fact that any $\mathcal{C} \in \mathcal{J}$ is finite dimensional, it is easy to infer that the measures $\mu_{\mathcal{C}}$ have finite supports.

By taking into account ([28], Corollary 1 to Lemma 3.2, Lemma 3.3, and Lemma 3.6), we infer that for any $a_1, a_2, \ldots, a_n \in A, n \geq 1$, we have

$$\int_{E_0(A)} \lambda_A(a_1)\lambda_A(a_2)\ldots\lambda_A(a_n)d\mu_{\mathcal{C}} = \int_{E_0(A)} \|\cdot\|\lambda_A(a_1)\ldots\lambda_A(a_n)d\mu_{\mathcal{C}}$$

$$= \left(K_{\mu_{\mathcal{C}}}\left(\lambda_A(a_1)\lambda_A(a_2)\ldots\lambda_A(a_n)\right)\xi^0_{b(\mu)} \middle| \xi^0_{b(\mu)} \right)$$

$$= \left(K_{\mu_{\mathcal{C}}}\left(\lambda_A(a_1)\right) K_{\mu_{\mathcal{C}}}\left(\lambda_A(a_2)\right)\ldots K_{\mu_{\mathcal{C}}}\left(\lambda_A(a_n)\right)\xi^0_{b(\mu)} \middle| \xi^0_{b(\mu)} \right)$$

$$= \left(e_{\mathcal{C}}\pi_{b(\mu)}(a_1)e_{\mathcal{C}}\pi_{b(\mu)}(a_2)e_{\mathcal{C}}\ldots e_{\mathcal{C}}\pi_{b(\mu)}(a_n)e_{\mathcal{C}}\xi^0_{b(\mu)} \middle| \xi^0_{b(\mu)} \right)$$

$$\to \left(e_\mu\pi_{b(\mu)}(a_1)e_\mu\pi_{b(\mu)}(a_2)e_\mu\ldots e_\mu\pi_{b(\mu)}(a_n)e_\mu\xi^0_{b(\mu)} \middle| \xi^0_{b(\mu)} \right)$$

$$= \int_{E_0(A)} \lambda_A(a_1)\lambda_A(a_2)\ldots\lambda_A(a_n)d\mu.$$

Since we have $1 = \|\mu_C\| = \mu_C(E(A))$, from the Stone–Weierstrass Theorem we now infer that $\lim_C \mu_C = \mu$ in the w^*-topology on $E_0(A)$. The Lemma is proved.

Corollary. *Let μ be a subcentral Radon probability measure on $E_0(A)$, whose barycenter is a state. Then there exists an increasing net (μ_C) in the set $\Omega(E_0(A); b(\mu))$, consisting of subcentral Radon probability measures on $E_0(A)$, and having finite supports, such that*

(a) $\sup\{\mu_C(f); \mathcal{C}\} = \mu(f), \ f \in S(E_0(A))$;
(b) $\lim_C \mu_C = \mu$ in $\sigma(C(E_0(A))^*; C(E_0(A)))$.

Proof: This is an immediate consequence of the proof of the preceding Lemma: indeed, if μ is subcentral, then $\mathcal{C}_\mu$ is contained in the center of $\pi_{b(\mu)}(A)'$, and, therefore, all the von Neumann algebras $\mathcal{C} \in \mathcal{J}$ are also contained in the center of $\pi_{b(\mu)}(A)'$; hence, the measures μ_C are subcentral.

The following Lemma is a slight improvement of a Lemma of C. F. Skau (see [25], Lemma 12; [28], Lemma 3.7; see, also, [41], proof of Lemma 3.19, where only subcentral measures are considered).

Lemma 4. *Let $\mu, \nu \in \mathcal{M}_+^1(E_0(A))$ be such that $\mu \prec \nu$. Then, for any subset*

$$\{\phi_1, \ldots, \phi_m\} \subset L^\infty(\mu)_1^+,$$

such that $\sum_{i=1}^n \phi_i = 1$, there exists a subset

$$\{\psi_1, \ldots, \psi_m\} \subset L^\infty(\nu)_1^+,$$

such that $\sum_{i=1}^m \psi_i = 1$ and $K_\mu(\psi_i) = K_\nu(\psi_i), 1 \leq i \leq m$.

Proof: Let us define $\mu_i = \phi_i \mu, 1 \leq i \leq m$. Then we have $\mu = \sum_{i=1}^m \mu_i$. From the Cartier–Fell–Meyer Theorem (see [28], Theroem 1.6), we infer that there exist positive Radon measures $\nu_i, 1 \leq i \leq m$, on $E_0(A)$, such that $\nu = \sum_{i=1}^m \nu_i$ and $\mu_i \sim \nu_i, 1 \leq i \leq m$. We infer that there exist functions $\psi_i \in L^\infty(\nu)_1^+$, such that $\nu_i = \psi_i \nu, 1 \leq i \leq m$, and, of course, we have $\sum_{i=1}^m \psi_i = 1 \pmod{\nu}$. We then have

$$
\begin{aligned}
(K_\nu(\psi_i)\pi_{f_0}(a)\xi_{f_0}^0 | \xi_{f_0}^0) &= \int_{E_0(A)} \psi_i \lambda_A(a) d\nu = \int_{E_0(A)} \lambda_A(a) d\nu_i \\
&= \int_{E_0(A)} \lambda_A(a) d\mu_i = \int_{E_0(A)} \phi_i \lambda_A(a) d\mu \\
&= (K_\mu(\phi_i)\pi_{f_0}(a)\xi_{f_0}^0 | \xi_{f_0}^0), \ a \in A,
\end{aligned}
$$

(1)

for any $i \in \{1, 2, \ldots, m\}$, where we have denoted $f_0 \stackrel{\text{def}}{=} b(\mu) = b(\nu)$, taking into account the fact that $\mu \prec \nu \Rightarrow \mu \sim \nu$. From (1) we infer that $K_\nu(\phi_i) = K_\mu(\psi_i), 1 \leq i \leq m$. The Lemma is proved.

The following lemma essentially belongs to W. Wils (see [41], proof of Proposition 3.20; and also [25], Lemma 24).

Lemma 5. *Let* $\mu = \sum_{i=1}^m c_i \epsilon_{f_i}, \nu = \sum_{j=1}^n d_j \epsilon_{g_j}$ *be two measures on* $E_0(A)$, *with finite supports, such that*

(a) $f_i \in E(A), 1 \leq i \leq m; g_j \in E(A), 1 \leq j \leq n;$
(b) $c_i \geq 0, 1 \leq i \leq m; d_j \geq 0, 1 \leq j \leq n; \sum_{i=1}^m c_i = \sum_{j=1}^n d_j = 1;$
(c) μ *is subcentral;*
(d) $b(\mu) = b(\nu)$.

Then there exists the least upper bound of μ *and* ν *in* $\mathcal{M}^1_+(E_0(A); f_0)$, *where* $f_0 \stackrel{\text{def}}{=} b(\mu) = b(\nu)$.

Proof: We can assume that $f_i \neq f_j$ and $g_i \neq g_j$ for $i \neq j$; and also that $c_i > 0, 1 \leq i \leq m$; and $d_j > 0, 1 \leq j \leq n$.

By hypothesis, the von Neumann algebra $\mathcal{C}_\mu$, corresponding to μ, is contained in $\pi_{f_0}(A)'' \cap \pi_{f_0}(A)'$.

Let $C_i \stackrel{\text{def}}{=} K_\mu(\chi_{\{f_i\}}), 1 \leq i \leq m$, and $D_j \stackrel{\text{def}}{=} K_\nu(\chi_{\{g_j\}}), 1 \leq j \leq n$. Then we have

(i) $C_i \in \pi_{f_0}(A)'' \cap \pi_f(A)', 1 \leq i \leq m;$
(ii) $C_i, 1 \leq i \leq m$, is a central projection and $\sum_{i=1}^m C_i = 1;$
(iii) $D_j \in \pi_{f_0}(A)', 1 \leq j \leq n;$
(iv) $D_j \geq 0, 1 \leq j \leq n$; and $\sum_{j=1}^n D_j = 1$.

Of course, we have

$$c_i f_i(a) = \int_{E_0(A)} \chi_{\{f_i\}} \lambda_A(a) d\mu = (C_i \pi_{f_0}(a) \xi_{f_0}^0 | \xi_{f_0}^0), 1 \leq i \leq m,$$

and

$$d_j g_j(a) = \int_{E_0(A)} \chi_{\{g_j\}} \lambda_A(a) d\mu = (D_j \pi_{f_0}(a) \xi_{f_0}^0 | \xi_{f_0}^0), 1 \leq j \leq n,$$

for any $a \in A$. We shall define the positive linear functional $r'_{ij} \in E_0(A)$ by

$$r'_{ij}(a) = (C_i D_j \pi_{f_0}(a) \xi_{f_0}^0 | \xi_{f_0}^0), \quad a \in A.$$

117

(Since $C_i D_j = D_j C_i$, we have $C_i D_j \geq 0$, and therefore, $r'_{ij} \geq 0, 1 \leq i \leq m, 1 \leq j \leq n$). If $r'_{ij} \neq 0$, define $r_{ij} = \|r'_{ij}\|^{-1} r'_{ij}$, if $r'_{ij} = 0$, define $r_{ij} = 0$. It is obvious that for

$$\theta \overset{\text{def}}{=} \sum_{i,j} \|r'_{ij}\| \epsilon_{r_{ij}},$$

we have that θ is a simple measure in $\mathcal{M}^1_+(E_0(A); f_0)$ and

$$c_i f_i = \sum_{j=1}^{n} \|r'_{ij}\| r_{ij}, 1 \leq i \leq m;$$

$$d_j g_j = \sum_{i=1}^{m} \|r'_{ij}\| r_{ij}, 1 \leq j \leq n.$$

By simple convexity arguments, we have that $\mu \prec \theta$ and $\nu \prec \theta$.

Conversely, let $\tau \in \mathcal{M}^1_+(E_0(A); f_0)$ be such that $\mu \prec \tau$ and $\nu \prec \tau$. We shall prove that $\theta \prec \tau$.

From Lemma 4 we infer that there exist functions $\phi_i \in L^\infty(\tau)^+_1$, such that $C_i = K_\tau(\phi_i), 1 \leq i \leq m$; and also, there exist functions $\psi_j \in L^\infty(\tau)^+_1$, such that $D_j = D_\tau(\psi_j), 1 \leq j \leq n$; moreover, we can assume that

$$\sum_{i=1}^{m} \phi_i = \sum_{j=1}^{n} \psi_j = 1 \ (\text{mod } \tau).$$

Of course, any decomposition $\theta = \sum_{k=1} \theta_k$ of θ is of the form

$$\theta_k = \sum_{i,j} t_{ijk} \|r'_{ij}\| \epsilon_{r_{ij}},$$

where $0 \leq t_{ijk}, 1 \leq k \leq l$; and $\sum_{k=1}^{l} t_{ijk} = 1, 1 \leq i \leq m, 1 \leq j \leq n$.

If we define

$$\tau_k = \left(\sum_{i,j} t_{ijk} \phi_i \psi_j \right) \tau, \quad 1 \leq k \leq l,$$

we shall have

(1) $$\tau = \sum_{k=1}^{l} \tau_k \text{ and } \tau_k \sim \theta_k, \ 1 \leq k \leq l.$$

With the Cartier–Fell-Meyer Theorem we shall infer that $\theta \prec \tau$.

118

Indeed, we have

$$\sum_{k=1}^{l} \tau_k = \left(\sum_{i,j,k} t_{ijk}\phi_i\psi_j\right) = \left(\sum_{i,j}\left(\sum_{k=1}^{l} t_{ijk}\right)\phi_i\psi_j\right)$$

$$= \left(\sum_{i,j}\phi_i\psi_j\right) = \left(\sum_{i=1}^{m}\phi_i\right)\left(\sum_{j=1}^{n}\psi_j\right)\tau = \tau,$$

and the equality in (1) is proved.

For the second part of the assertion (1), we shall first prove that

$$(2) \qquad K_\tau(\phi_i\psi_j) = K_\tau(\phi_i)K_\tau(\psi_j), \quad 1 \le i \le m, 1 \le j \le n.$$

Indeed, from $0 \le \phi_i\psi_j \le \phi_i, \psi_j$, we infer that

$$0 \le K_\tau(\phi_i\psi_j) \le K_\tau(\phi_i), K_\tau(\psi_j), \quad 1 \le i \le m, 1 \le j \le n.$$

Since $K_\tau(\phi_i)$ is a projection, we infer that

$$K_\tau(\phi_i\psi_j) = K_\tau(\phi_i)K_\tau(\phi_i\psi_j) = K_\tau(\phi_i\psi_j)K_\tau(\phi_i);$$

whereas from the fact that $K_\tau(\phi_i)$ is a central projection, we infer that

$$(3) \qquad K_\tau(\phi_i\psi_j) \le K_\tau(\phi_i)K_\tau(\psi_j).$$

By replacing ϕ_i with $1 - \phi_i$, we similarly get that

$$K_\tau((1-\phi_i)\psi_j) \le K_\tau(1-\phi_i)K_\tau(\psi_j),$$

and this implies that

$$(4) \qquad K_\tau(\phi_i)K_\tau(\psi_j) \le K_\tau(\phi_i\psi_j).$$

From (3) and (4) we get the desired result (1).

Let us now remark that, by virtue of (2), we have

$$\int_{E_0(A)} \lambda_A(a)d\tau_k = \int_{E_0(A)} \left(\sum_{i,j} t_{ijk}\phi_i\psi_j \right) \lambda_A(a)d\tau$$

$$= \left(\sum_{i,j} t_{ijk}K_\tau(\phi_i\psi_j)\pi_{f_0}(a)\xi_{f_0}^0 \,\middle|\, \xi_{f_0}^0 \right)$$

$$= \sum_{i,j} t_{ijk} \left(K_\tau(\phi_i)K_\tau(\psi_j)\pi_{f_0}(a)\xi_{f_0}^0 \,\middle|\, \xi_{f_0}^0 \right)$$

$$= \sum_{i,j} t_{ijk} \left(C_i D_j \pi_{f_0}(a)\xi_{f_0}^0 \,\middle|\, \xi_{f_0}^0 \right)$$

$$= \sum_{i,j} t_{ijk} r'_{ij}(a)$$

$$= \sum_{i,j} t_{ijk} \left\| r'_{ij} \right\| r_{ij}(a)$$

$$= \sum_{i,j} t_{ijk} \left\| r'_{ij} \right\| \lambda_A(a)(r_{ij})$$

$$= \int_{E_0(A)} \lambda_A(a)d\theta_k, \ 1 \le k \le l, a \in A.$$

We also have that

$$\tau_k(1) = \int_{E_0(A)} \left(\sum_{i,j} t_{ijk}\phi_i\psi_j \right) d\tau = \int_{E_0(A)} \left(\sum t_{ijk}\phi_i\psi_j \right) \| \cdot \| d\tau$$

$$= \left(K_\tau \left(\sum_{i,j} t_{ijk}\phi_i\psi_j \right) \xi_{f_0}^0 | \xi_{f_0}^0 \right) = \sum_{i,j} t_{ijk} \left(C_i D_j \xi_{f_0}^0 | \xi_{f_0}^0 \right) = \theta_k(1),$$

for $1 \le k \le l$, by taking into account ([28], Corollary 1 to Lemma 3.2 and Lemma 3.3). From (5) and (6) we infer that $\tau_k \sim \theta_k, 1 \le k \le l$, and so the second assertion in (1) is proved, and also the Lemma.

We shall denote $\theta = \mu \vee \tau$.

The following Theorem belongs to W. Wils (see [41], Proposition 3.20, where it is stated under more general conditions). It is stated as Lemma 25 in [25], where it is given for C^*-algebras with a unit element. Here we shall drop this assumption.

Theorem 2. *For any state $f_0 \in E(A)$, any measure $\mu \in \mathcal{M}_+^1(E_0(A); f_0)$ and any subcentral measure $\nu \in \mathcal{M}_+^1(E_0(A); f_0)$ there exists the least upper bound of μ and ν in $\mathcal{M}_+^1(E_0(A); f_0)$, with respect to the Choquet–Meyer order relation.*

Proof: Let $(\mu_{\mathcal{L}})$ be an increasing net in $\mathcal{F}_+^1(E_0(A); f_0)$, as given by Lemma 2, for the measure μ. Let $(\nu_{\mathcal{C}})$ be an increasing net of simple subcentral measures, as given by the Corollary to Lemma 3, for the measure ν. Let us define $\theta_{\mathcal{L},\mathcal{C}} = \mu_{\mathcal{L}} \vee \nu_{\mathcal{C}}$, as given by Lemma 5. Since we have that $\theta_{\mathcal{L},\mathcal{C}} \in \mathcal{M}_+^1(E_0(A); f_0)$, and since this space is w^*-compact, we can select a subnet of $(\theta_{\mathcal{L},\mathcal{C}})$, which converges to $\theta \in \mathcal{M}_+^1(E_0(A); f_0)$. Since for any $\phi \in S(E_0(A))$ we have

$$\int_{E_0(A)} \phi \, d\mu_{\mathcal{L}} \leq \int_{E_0(A)} \phi \, d\theta_{\mathcal{L},\mathcal{C}}, \int_{E_0(A)} \phi d\nu_{\mathcal{C}} \leq \int_{E_0(A)} \phi \, d\theta_{\mathcal{L},\mathcal{C}}, \forall \mathcal{L}, \mathcal{C},$$

we infer that

$$\int_{E_0(A)} \phi \, d\mu \leq \int_{E_0(A)} \phi \, d\theta, \int_{E_0(A)} \phi \, d\nu \leq \int_{E_0(A)} \phi \, d\theta, \phi \in S(E_0(A)),$$

and this shows that $\nu \prec \theta, \nu \prec \theta$.

Let now $\tau \in \mathcal{M}_+^1(E_0(A); f_0)$ be such that $\mu \prec \tau$ and $\nu \prec \tau$. Then we have $\mu_{\mathcal{L}} \prec \tau, \nu_{\mathcal{C}} \prec \tau$ for any $\mathcal{L}$ and $\mathcal{C}$; therefore, by Lemma 5 we have $\theta_{\mathcal{L},\mathcal{C}} \prec \tau$ and this implies that $\theta \prec \tau$. It follows that $\theta = \mu \vee \nu$.

Remark. Since the preceding argument works for any converging subnet of $(\theta_{\mathcal{L},\mathcal{C}})$, and since the least upper bound, when it exists, is unique, we infer that we actually have $\theta = \lim \theta_{\mathcal{L},\mathcal{C}}$. Moreover, since

$$\mathcal{C}_1 \leq \mathcal{C}_2 \text{ and } \mathcal{L}_1 \leq \mathcal{L}_2$$

implies that

$$\theta_{\mathcal{L}_1,\mathcal{C}_1} \prec \theta_{\mathcal{L}_2,\mathcal{C}_2},$$

we see that we have

$$\sup \left\{ \int_{E_0(A)} \phi \, d\theta_{\mathcal{L},\mathcal{C}}; \mathcal{L}, \mathcal{C} \right\} = \int_{E_0(A)} \phi \, d\theta, \phi \in S(E_0(A)),$$

and so we have $\sup \{\theta_{\mathcal{L},\mathcal{C}}; \mathcal{L}, \mathcal{C}\} = \theta$ in $\mathcal{M}_+^1(E_0(A); f_0)$.

The following Theorem was first proved by W. Wils (see [40]) for the case of C^*-algebras having a unit element (see, also, [25], Theorem 26).

Theorem 3. *Let $f_0 \in E(A)$ be a state of A and let $\mu \in \mathcal{M}_+^1(E_0(A); f_0)$ be any subcentral measure. Then we have $\mu \prec \nu$ for any Choquet maximal measure $\nu \in \mathcal{M}_+^1(E_0(A); f_0)$.*

Proof: By Theorem 2, the least upper bound $\mu \vee \nu$ exists in $\mathcal{M}_+^1(E_0(A); f_0)$. Since we have $\nu \prec \mu \vee \nu$, and since ν is maximal, we infer that $\nu = \mu \vee \nu$ and therefore, $\mu \prec \nu$. The Theorem is proved.

Proof of Theorem 1: Let $f_0 \in E(A)$ and let μ_{f_0} be the central measure corresponding to f_0. Then we have $\mu_f \prec \nu$, for any Choquet maximal measure $\nu \in \mathcal{M}_+^1(E_0(A); f_0)$, by virtue of Theorem 3.

Conversely, let $\mu \in \mathcal{M}_+^1(E_0(A); f_0)$ be a measure, such that $\mu \prec \nu$, for any Choquet maximal measure $\nu \in \mathcal{M}_+^1(E_0(A); f_0)$. By virtue of Henrichs' Theorem (see [18]; [28], Theorem 3.10), we shall have $\mu \prec \nu$, for any *maximal orthogonal* measure $\nu \in \Omega(E_0(A); f_0)$. By ([28], Lemma 3.7), we infer that

$$K_\mu \left(L^\infty(\mu)_1^+ \right) \subset K_\nu \left(L^\infty(\nu)_1^+ \right),$$

for any maximal orthogonal measure ν, and, therefore,

$$
(1) \qquad
\begin{aligned}
K_\mu \left(L^\infty(\mu)_1^+ \right) \subset \cap_\nu K_\nu \left(L^\infty(\nu)_1^+ \right) &= \left[\pi_{f_0}(A)'' \cap \pi_{f_0}(A)' \right]_1^+ \\
&= K_{\mu_{f_0}} \left(L^\infty(\mu_{f_0})_1^+ \right),
\end{aligned}
$$

where the intersection is taken over all maximal orthogonal measures. Since μ_{f_0} is simplicial (see [28], Corollary 1 to Theorem 3.1), from (1) and from ([28], Lemma 3.7, ii)) we infer that $\mu \prec \mu_{f_0}$. The Theorem is proved.

III. The following theorem essentially belongs to W. Wils (see [41]). We shall denote by $\mathcal{Z}_+^1(E_0(A); f_0)$ the set of all subcentral Radon probability measures on $E_0(A)$, whose barycenter is at f_0.

Theorem 4. *For any state $f_0 \in E(A)$, the set $\mathcal{Z}_+^1(E_0(A); f_0)$ is a complete lattice with respect to the Choquet–Meyer order relation.*

Proof: The mapping $\mathcal{Z}_+^1(E_0(A); f_0) \ni \mu \mapsto \mathcal{C}_\mu \subset \pi_{f_0}(A)' \cap \pi_{f_0}(A)''$ is an order isomorphism between $\mathcal{Z}_+^1(E_0(A); f_0)$ and the set of all von Neumann subalgebras of the center of $\pi_{f_0}(A)'$ (see [28], Theorem 3.4).

As usual, for any operator $a \in \mathcal{L}(H)$ we shall denote by $r(a)$, respectively $l(a)$, its *right*, respectively *left support*. If $a^* = a$, then $l(a) = r(a)$, and this projection is called the *support* of a and it is denoted by $s(a)$.

Theorem 5. *Let $\mu \in \mathcal{M}_+^1(E_0(A))$ and $\phi \in L^\infty(\mu), \phi \geq 0$. Then*

(a) *if μ is a subcentral measure, then $\phi\mu$ is a subcentral measure;*

(b) *if μ is a central measure, then $\phi\mu$ is a central measure.*

Proof: It is known that if μ is an orthogonal measure, then $\phi\mu$ is also orthogonal (see [28], Proposition 3.16).

(a) Let $f_0 = r(\mu)$. Then the mapping K_μ maps $L^\infty(\mu)$ into the center $\pi_{f_0}(A)' \cap \pi_{f_0}(A)''$ of $\pi_{f_0}(A)'$, and

$$\int_{E_0(A)} \psi \lambda_A(a) d\mu = \left(K_\mu(\psi)\pi_{f_0}(a)\xi_{f_0}^0 | \xi_{f_0}^0 \right), \quad a \in A, \psi \in L^\infty(\mu).$$

Since we have

$$
\begin{aligned}
(1) \qquad \int_{E_0(A)} \psi \lambda_A(a) d(\psi\mu) &= \int_{E_0(A)} \psi \phi \lambda_A(a)\, d\mu \\
&= \left(K_\mu([\psi])\pi_{f_0}(a) K_\mu([\phi])^{\frac{1}{2}} \xi_{f_0}^0 | K_\mu([\phi])^{\frac{1}{2}} \xi_{f_0}^0 \right)
\end{aligned}
$$

for any $a \in A$ and $\psi \in L^\infty(\psi\mu)$, we infer that $H_{r(\phi\mu)}$ can be identified with $s(K_\mu(\phi))H_{r(\mu)}$, $\xi_{r(\phi\mu)}^0$ can be identified with $K_\mu(\phi)^{\frac{1}{2}}\xi_{f_0}^0$, whereas $\pi_{r(\phi\mu)}$ can be identified with the representation

$$A \ni a \mapsto \pi_{f_0}(a)s(K_\mu(\phi)).$$

From (1) we infer that

$$(2) \qquad K_{\phi\mu}([\psi]) = K_\mu([\psi])s(K_\mu([\phi])), \quad \psi \in \mathcal{L}^\infty(\mathcal{B}(E_0(A))).$$

(We can always use bounded Borel measurable representatives of elements in $L^\infty(\mu)$, etc.) Of course, we have used the fact that $s(K_\mu([\phi])) \in \mathcal{C}_\mu$.

From (2) we infer that $K_{\phi\mu}$ maps $L^\infty(\phi\mu)$ into the center of $\pi_{r(\phi\mu)}(A)$; hence, $\phi\mu$ is subcentral.

(b) If μ is central, then K_μ maps $L^\infty(\mu)$ *onto* $\pi_{f_0}(A)' \cap \pi_{f_0}(A)''$, whence we infer that $K_{\phi\mu}$ maps $L^\infty(\phi\mu)$ *onto* $\pi_{g_0}(A)' \cap \pi_{g_0}(A)''$, where $g_0 \overset{\text{def}}{=} r(\phi\mu)$. It follows that $\phi\mu$ is central. The Theorem is proved.

IV. The following Theorem characterizes the central measures. It is implicitly contained in the proof following Definition 3.5.1 in [24], where the C^*-algebra A is assumed to possess the unit element. Since at the head of Section 3.5, p. 146, in [24], A is assumed, moreover, to be separable, a superficial reading might leave the impression that Sakai's characterization is established for the separable case only. In fact, it is easy to see that the argument following Definition 3.5.1 in [24] does not require the separability of A, and so the following Theorem is only the extension of Sakai's result to the slightly more general case of a C^*-algebra A, possibly not possessing the unit element.

Theorem 6. *Let A be any C^*-algebra, $f_0 \in E(A)$ a state of A, and denote by π_{f_0}'' : $A^{**} \to \mathcal{L}(H_{f_0})$ the canonical normal extension of π_{f_0} to A^{**}. Then a measure $\mu \in \mathcal{M}_+^1(E_0(A); f_0)$ is the central measure μ_{f_0} if, and only if, there exists a $*$-homomorphism $\Phi : Z(A^{**}) \to L^\infty(\mu)$, of the center $Z(A^{**})$ of A^{**}, onto $L^\infty(\mu)$, such that*

(1)
$$f_0(za) = \int_{E_0(A)} \Phi(z)\lambda_A(a) \, d\mu, \quad a \in A, z \in Z(A^{**}).$$

In this case, Φ is unique.

(Of course, in (1), for $\Phi(z)$ one should take a function representative of the class $\Phi(z) \in L^\infty(\mu)$; it can always be chosen to be a bounded Baire measurable function on $E_0(A)$, whereas with the help of Lifting Theory, one can even obtain a $*$-homomorphism into $\mathcal{L}^\infty(E_0(A); \mu)$.)

Proof: Let us assume that $\mu = \mu_{f_0}$. Then, by its definition, the mapping K_μ maps isomorphically the C^*-algebra $L^\infty(\mu)$ onto the center of $\pi_{f_0}(A)'$ (hence, also of $\pi_{f_0}(A)''$). On the other hand, the restriction $\pi_f'' | Z(A^{**})$ is a normal $*$-homomorphism of $Z(A^{**})$ onto the center of $\pi_{f_0}(A)''$, and we have

$$f_0(za) = \left(\pi_{f_0}(a)\pi_{f_0}''(z)\xi_{f_0}^0 | \xi_{f_0}^0\right), \quad a \in A, z \in Z(A^{**}).$$

The mapping $\Phi_{f_0} = K_{\mu_{f_0}}^{-1} \circ (\pi_{f_0}'' | Z(A^{**}))$ has then the required property (1) (see [28], Lemma 3.3, Proposition 3.1, ii) and Corollary 1 to Theorem 3.1).

Conversely, let us assume that there exists a surjective $*$-homomorphism $\Phi : Z(A^{**}) \to L^\infty(\mu)$, such that equality (1) be satisfied.

(a) For any $z \in Z(A^{**})$ we shall have

$$f_0(za) = \left(\pi_{f_0}(a)\pi_{f_0}''(z)\xi_{f_0}^0 | \xi_{f_0}^0\right) = \int_{E_0(A)} \Phi(z)\lambda_A(a) \, d\mu$$
$$= (K_\mu(\Phi(z))\pi_{f_0}(a)\xi_{f_0}^0 | \xi_{f_0}^0), \quad a \in A,$$

and this implies that

(2)
$$K_\mu(\Phi(z)) = \pi_{f_0}''(z), \quad z \in Z(A^{**}).$$

(b) For any $\phi_1, \phi_2 \in L^\infty(\mu)$ there exists $z_1, z_2 \in Z(A^{**})$, such that $\Phi(z_i) = \phi_i, i = 1, 2$. We shall then have

$$K_\mu(\phi_1\phi_2) = K_\mu(\Phi(z_1)\Phi(z_2)) = K_\mu(\Phi(z_1 z_2)) = \pi_{f_0}''(z_1 z_2)$$
$$= \pi_{f_0}''(z_1)\pi_{f_0}''(z_2) = K_\mu(\Phi(z_1))K_\mu(\Phi(z_2)) = K_\mu(\phi_1)K_\mu(\phi_2),$$

and this shows that μ is orthogonal. From (2) we infer that μ is central, whereas $\|b(\mu)\| = 1$ implies that K_μ is injective. From (2) we infer that

$$\Phi(z) = \left(K_\mu^{-1} \circ \pi''_{f_0} \right)(z), \quad z \in Z(A^{**}),$$

and so the uniqueness of Φ is established. The Theorem is proved.

We shall denote by Φ_{f_0} the $*$-homomorphism corresponding to the state $f_0 \in E(A)$, and whose existence and uniqueness is established above.

Lemma 6. *For any two positive linear functionals $f_1, f_2 \in A^*_+$ consider the relations:*

(a) $f_1 \overset{\downarrow}{\circ} f_2$;

(b) $\|f_1 - f_2\| = \|f_1\| + \|f_2\|$;

(c) $f_1 \perp f_2$;

(d) $f_1 \neq f_2$, *unless* $f_1 = f_2 = 0$.

Then (a) $\Rightarrow$ (b) $\Rightarrow$ (c) $\Rightarrow$ (d), and no implication here can be reversed.

Proof: Indeed, there exists a projection $e_0 \in Z(A^{**})$, such that $f_1(a) = f_1(ae_0), f_2(a) = f_2(a(1 - e_0)), a \in A$. If we denote $f = f_1 + f_2$, then we have

$$f(e_0 a) = f_1(a), \quad a \in A,$$
$$f((1 - e_0)a) = f_2(a), \quad a \in A,$$

and therefore,

$$(f_1 - f_2)(a) = f((2e_0 - 1)a), \quad a \in A,$$

whence we get that

$$\|f_1 - f_2\| = \|f \cdot (2e_0 - 1)\| = \|f\| = \|f_1\| + \|f_2\|,$$

because $2e_0 - 1$ is unitary. Thus, the implication (a) $\Rightarrow$ (b) is proved.

(b) $\Rightarrow$ (c). Indeed, assume that $0 \le f \le f_1$ and $0 \le f \le f_2$. Then we have $f_i = f + (f_i - f), i = 1, 2$, and, therefore

$$\|f_1\| + \|f_2\| = \|f_1 - f_2\| = \|(f_1 - f) - (f_2 - f)\|$$
$$\le \|f_1 - f\| + \|f_2 - f\| = \|f_1\| + \|f_2\| - 2\|f\|,$$

and this implies that $f = 0$. Hence, $f_1 \perp f_2$.

(c) $\Rightarrow$ (d). If $f_1 \perp f_2$ and $f_1 = f_2$, then, obviously, $f_1 = f_2 = 0$. Simple examples can be found to show that the preceding implications cannot be reversed, in general.

Lemma 7. Let $f_1, f_2 \in E(A)$ be two states, such that

$$2 = \|f_1 - f_2\|$$

and let $\mu_i \in \mathcal{M}^1_+(E_0(A); f_i), i = 1, 2$. Then the measures μ_1 and μ_2 are mutually singular.

Proof: Indeed, by ([12], Proposition 12.3.1) the supports $s(f_1)$, resp. $s(f_2)$, of f_1, resp. f_2, in A^{**} are orthogonal: $s(f_1)s(f_2) = 0$. Since $s(f_1)$ and $s(f_2)$ are countably decomposable in A^*, by ([42], Chapter I, Lemma 1.2) there exists a Baire element $b \in \mathcal{B}_0(A) \subset A^{**}$ (see [31]), such that $0 \le b \le 1$ and $f_1(b) = \|f_1\| = 1, f_2(b) = 0$. If we denote $e = \lim_{n \to \infty} b^n$, then e is a Baire projection and we have

$$f_1(e) = 1, \quad f_2(e) = 0.$$

Since the barycentric calculus holds for Baire elements over A (see [31], Theorem 3), we infer that

$$(1) \qquad \int_{E_0(A)} \lambda_A(e) \, d\mu_1 = 1, \quad \int_{E_0(A)} \lambda_A(e) \, d\mu_2 = 0.$$

From (1) we infer that

$$\mu_1 \left(\{ f \in E_0(A); f(e) = 1 \} \right) = 1$$

and

$$\mu_2 \left(\{ f \in E_0(A); f(e) = 1 \} \right) = 0.$$

Hence, μ_1 and μ_2 are mutually singular.

Theorem 7. Let $f_1, f_2 \in E(A)$ be two states, such that $f_1 \perp f_2$. Let $t \in [0, 1]$ and define $f = tf_1 + (1-t)f_2$. Then

(a) $\mu = t\mu_{f_1} + (1-t)\mu_{f_2}$ is the central measure corresponding to f;
(b) $\Phi_f = \Phi_{f_1} + \Phi_{f_2}$.

Proof: (a). By Lemma 6 and by Godement's Theorem (see [25], p. 281; [17]) we can identify $\pi_f : A \to \mathcal{L}(H_f)$ with $\pi = \pi_{f_1} \oplus \pi_{f_2} : A \to \mathcal{L}(H_{f_1} \oplus H_{f_2})$, with the associated cyclic vector

$$\xi_0 = \left(t^{\frac{1}{2}} \xi^0_{f_1}, (1-t)^{\frac{1}{2}} \xi^0_{f_2} \right).$$

We shall assume from now on that $0 < t < 1$. Otherwise, the assertion is trivial.

For any bounded Borel measurable complex function ϕ on $E_0(A)$ we shall have

$$\int_{E_0(A)} \phi \lambda_A(a) \, d(t\mu_{f_1} + (1-t)\mu_{f_2}) = t \int_{E_0(A)} \phi \lambda_A(a) \, d\mu_{f_1}$$

$$+ (1-t) \int_{E_0(A)} \phi \lambda_A(a) \, d\mu_{f_2}$$

$$= t \left(K_{\mu_{f_1}}(\phi) \pi_{f_1}(a) \xi^0_{f_1} | \xi^0_{f_1} \right)$$

$$+ (1-t) \left(K_{\mu_{f_2}}(\phi) \pi_{f_2}(a) \xi^0_{f_2} | \xi^0_{f_2} \right)$$

$$= \left(\left(K_{\mu_{f_1}}(\phi) \pi_{f_1}(a) \oplus K_{\mu_{f_2}}(a) \right) \xi_0 | \xi_0 \right)$$

$$= \left(\left(K_{\mu_{f_1}}(\phi) \oplus K_{\mu_{f_2}}(\phi) \right) \pi_f(a) \xi_0 | \xi_0 \right), \quad a \in A,$$

and this shows that

(1) $\qquad K_{\mu_{f_1}}(\phi) \oplus K_{\mu_{f_2}}(\phi) = K_{t\mu_{f_1} + (1-t)\mu_{f_2}}(\phi), \phi \in \mathcal{L}^\infty(E_0(A), \mathcal{B}(E_0(A))).$

Remark. This equality is established only on the basis of the assumption that $f_1 \perp f_2$. Under this assumption it is easy to infer that μ is subcentral.

Let $e_i : H_{f_1} \oplus H_{f_2} \to H_{f_i}, i = 1, 2$, be the canonical projections. Of course, we have $e_1, e_2 \in \pi(A)'$. From $f_1 \perp f_2$ we now infer that

(2) $$\pi(A)' = \pi_{f_1}(A)' \oplus \pi_{f_2}(A)'$$

and

(3) $$\pi(A)'' = \pi_{f_1}(A)'' \oplus \pi_{f_2}(A)''$$

(see [22], Theorem 3.8.11). We infer that

$$e_1, e_2 \in \pi(A)' \cap \pi(A)''.$$

From Lemma 6 and Lemma 7 we now infer that there exists a Borel measurable subset $M \subset E_0(A)$, such that

$$\mu_{f_1}(M) = 1, \mu_{f_2}(M) = 0.$$

We infer that

(4) $$K_{\mu_{f_1}}(\chi_M) = 1_{H_{f_1}}, K_{\mu_{f_2}}(\chi_M) = 0.$$

From (1) we immediately infer that the measure $t\mu_{f_1} + (1-t)\mu_{f_2}$ is orthogonal; whereas from (2) and (3) it follows that this measure is subcentral. Moreover, we have

(5)
$$\pi(A)' \cap \pi(A)'' = \left(\pi_{f_1}(A)' \cap \pi_{f_1}(A)''\right) \oplus \left(\pi_{f_2}(A)' \cap \pi_{f_2}(A)''\right).$$

For any $\phi_1, \phi_2 \in \mathcal{L}^\infty(E_0(A); \mathcal{B}(E_0(A)))$, let us define

$$\phi = \phi_1 \chi_M + \phi_2 \chi_{CM}.$$

From (4) we infer that $K_{\mu_1}(\phi) = K_{\mu_1}(\phi_1), K_{\mu_2}(\phi) = K_{\mu_2}(\phi_2)$; hence, by taking into account (1), we infer that μ is central (we have denoted $\mu = t\mu_{f_1} + (1-t)\mu_{f_2}$). Of course, $b(\mu) = f$.

(b) On account of equalities (4), it is obvious that we can assume that

(6)
$$\Phi_{f_1}(z)\chi_M = \Phi_{f_1}(z), \Phi_{f_2}(z)\chi_{CM} = \Phi_{f_2}(z), \ z \in Z(A^{**}).$$

From the equalities (6) it is obvious that the mapping

$$Z(A^{**}) \ni z \mapsto \Phi_1(z) + \Phi_2(z) \in \mathcal{L}^\infty(\mu),$$

is a multiplicative $*$-homomorphism, easily shown to be surjective, onto $L^\infty(\mu)$. On the other hand, it is obvious that

$$f(za) = \int_{E_0(A)} (\Phi_1(z) + \Phi_2(z)) \lambda_A(a) \, d\mu, \quad a \in A, z \in Z(A^{**}).$$

From Theorem 6 we infer that $\Phi_f = \Phi_{f_1} + \Phi_{f_2}$, and the Theorem is proved.

Remark. Since the measures μ_{f_1} and μ_{f_2} are mutually singular, there is a canonical isomorphism (for $0 < t < 1$)

$$j : L^\infty(t\mu_{f_1} + (1-t)\mu_{f_2}) \to L^\infty(\mu_{f_1}) \oplus L^\infty(\mu_{f_2}).$$

Then part (b) of the Theorem is better expressed as follows

$$j \circ \Phi_f = \Phi_{f_1} \oplus \Phi_{f_2}.$$

§3. The Central Topology

In this section we shall introduce the *central topology* on the set $F(A)$ of the factorial states of any C^*-algebra A.

I. A subset $F \subset E_0(A)$ will be said to be *centrally extremal* (*Z-extremal*, for short), if the relations

$$f \in F \cap E(A), f = tf_1 + (1-t)f_2, 0 < t < 1, f_1 \downarrow_\circ f_2, f_1, f_2 \in E_0(A)$$

imply $f_1, f_2 \in F$.

Of course, any extremal subset of $E_0(A)$, and any orthogonally extremal subset of $E_0(A)$, are Z-extremal (see [28], p. 141, for the definition of the orthogonally extremal subsets; and also, the Remark below). Also, it is obvious that any subset of $F(A)$ is Z-extremal.

Theorem 8. *For any compact subset $F \subset E_0(A)$ the following are equivalent:*

(a) *F is Z-extremal;*
(b) *for any subcentral measure $\mu \in M^1_+(E_0(A))$, such that $b(\mu) \in F \cap E(A)$ we have $\mu(F) = 1$.*

Proof: (a) $\Rightarrow$ (b) By way of contradiction, we shall assume that μ is a subcentral measure in $M^1_+(E_0(A))$, such that $b(\mu) \in F \cap E(A)$, and $\mu(F) < 1$. Then we have $\mu \neq \epsilon_{b(\mu)}$ (the Dirac measure at $b(\mu)$), and therefore, supp μ has at least two points. If supp $\mu \subset F$, then $\mu(F) = 1$, and this contradicts the assumption. It follows that supp $\mu \not\subset F$. Let then $f_0 \in (\text{supp } \mu) \backslash F$, and let $K_1 \subset E_0(A)$ be a compact convex neighbourhood of f_0, such that

$$K_1 \cap F = \emptyset \text{ and } f_1 \in \text{int } K_1.$$

Of course, we have $\mu(K_1) > 0$. If $\mu(K_1) = 1$, then we would have $b(\mu) \in K_1$; hence, $b(\mu) \notin F$, a contradiction. Therefore, we have $0 < \mu(K_1) < 1$. Let us define

$$\mu_1 = \mu(K_1)^{-1} \chi_{K_1} \mu$$

and

$$\mu_2 = (1 - \mu(K_1))^{-1} \chi_{\complement K_1} \mu.$$

Since μ is subcentral, by Lemma 1 we have

$$b(\mu_1) \downarrow_\circ b(\mu_2)$$

and, of course, from $\mu = \mu(K_1)\mu_1 + (1 - \mu(K_1))\mu_2$, we get

$$b(\mu) = \mu(K_1)b(\mu_1) + (1 - \mu(K_1))b(\mu_2) \in F \cap E(A),$$

whence, by virtue of (a), we infer that $b(\mu_1), b(\mu_2) \in F$. Since it is obvious that $b(\mu_1) \in K_1$, we arrived at a contradiction.

(b) $\Rightarrow$ (a) Let $f_1, f_2 \in E_0(A)$ be such that $f_1 \overset{|}{\circ} f_2$ and assume that there exists a $t \in (0,1)$, such that

$$t f_1 + (1 - t) f_2 \in F \cap E(A);$$

then the measure $\mu = t\epsilon_{f_1} + (1-t)\epsilon_{f_2}$ is subcentral ($f_1 \overset{|}{\circ} f_2$ implies that $f_1 \perp f_2$; hence, μ is orthogonal. Apply then Godement's Theorem). Since

$$b(\mu) = t f_1 + (1 - t) f_2 \in F \cap E(A),$$

we infer that $\mu(F) = 1$, and, therefore, $f_1, f_2 \in F$. The Theorem is proved.

Remark. We can strengthen the concept of Z-extremality of a subset $F \subset E_0(A)$ as follows: we shall say that F is *measure Z-extremal* (or Z-extremal in measure) if:

(a) F is a universally measurable subset of $E_0(A)$;

(b) For any subcentral measure $\mu \in \mathcal{M}_+^1(E_0(A))$, such that $b(\mu) \in F \cap E(A)$ we have $\mu(F) = 1$.

We recall that in ([28], p. 141) we introduced the notion of an *orthogonally extremal* (*ω-extremal*, for short) subset $F \subset E_0(A)$; namely, F is said to be ω-extremal if $f_0 \in F \cap E(A), f_0 = t f_1 + (1-t) f_2, 0 < t < 1, f_1, f_2 \in E_0(A), f_1 \perp f_2$, implies that $f_1, f_2 \in F$. A subset $F \subset E_0(A)$ is *measure ω-extremal* if F is universally measurable and if for any orthogonal Radon probability measure μ on $E_0(A)$ we have the implication

$$b(\mu) \in F \cap E(A) \Rightarrow \mu(F) = 1.$$

In ([33], p. 8) we introduced the notion of a *maximally orthogonally extremal* subset $F \subset E_0(A)$ (*Ω-extremal*, for short), as being any *compact* subset $F \subset E_0(A)$, such that for any maximal orthogonal Radon probability measure μ on $E_0(A)$ one should have that

$$b(\mu) \in F \cap E(A) \Rightarrow \mu(F) = 1.$$

Of course, this notion can be extended a little, by considering *universally measurable* subsets $F \subset E_0(A)$, for which the same implictaion should hold; we obtain then the notion of a *measure Ω-extremal* subset.

130

It is easy to see that *any ω-extremal* subset $F \subset E_0(A)$ *is Z-extremal*; and also, *any measure ω-extremal subset is measure Z-extremal*.

We shall denote by $Z(E_0(A))$ the set of all compact Z-extremal subsets of $E_0(A)$. It is obvious that:

(i) any finite union of elements in $Z(E_0(A))$ is again in $Z(E_0(A))$;

(ii) any intersection of elements in $Z(E_0(A))$ is again in $Z(E_0(A))$;

(iii) for any $f_0 \in F(A)$ the set $\{f_0\}$ belongs to $Z(E_0(A))$.

We immediately infer that the set

$$\widehat{Z}(F(A)) = \{F \cap F(A); F \in Z(E_0(A))\}$$

is the set of all closed subsets of $F(A)$, for a topology on $F(A)$, which we shall call the *central topology* of $F(A)$.

It is obvious from (iii) that $F(A)$, endowed with the central topology, is a (T_1)-space.

II. Let $F \subset E_0(A)$ be any subset. We shall say that $f_0 \in F$ is a *Z-extremal* point of F if there is *no* decomposition of the form

$$f_0 = tf_1 + (1-t)f_2, 0 < t < 1, f_1 \overset{|}{\circ} f_2, f_1, f_2 \in F.$$

It is obvious that *any extremal point of F is Z-extremal* (with the exception of 0, if $0 \in F$!).

We shall denote by $ex_Z F$ the set of all Z-extremal points of F.

Theorem 9. (a) For any subset $F \subset E_0(A)$ we have $(ex\ F) \backslash \{0\} \subset ex_Z F$; (b) For any Z-extremal subset $F \subset E_0(A)$ we have

$$(ex_Z\ F) \cap E(A) = F \cap F(A).$$

Proof: (a) Obvious.

(b) Let $f_0 \in (ex_Z F) \cap E(A)$; then $f_0 \in F$ and $f_0 \in E(A)$. If $f_0 \notin F(A)$, then there exists a decomposition of the form

$$f_0 = tf_1 + (1-t)f_2,$$

where $t \in (0,1), f_1 \overset{|}{\circ} f_2$ and $f_1, f_2 \in E(A)$. It follows that $f_1, f_2 \in F$, from the Z-extremality of F, and this contradicts the Z-extremality of f_0.

Conversely, if $f_0 \in F \cap F(A)$, then it is obvious that $f_0 \in (ex_Z F) \cap E(A)$. The Theorem is proved.

Theorem 10. *For any compact Z-extremal subset $F \subset E_0(A)$, such that $F \cap E(A) \neq \emptyset$, we have also $F \cap F(A) \neq \emptyset$.*

Proof: By the Converse Milman Theorem, we have that $ex \; \overline{co}(F) \subset F$. If we had $\|f\| < 1$, for any $f \in ex \; \overline{co}(F)$, then from the Strict Minimum Principle (see [28], Theorem 1.2) we would infer that $\|f\| < 1$, for any $f \in \overline{co}(F)$; hence, $\|f\| < 1$, for any $f \in F$, a contradiction. We infer that there exists a $f_0 \in ex \; \overline{co}(F) \cap E(A)$. But then we have $f_0 \in (ex \; F) \cap E(A) \subset (ex_Z \; F) \cap E(A) = F \cap F(A)$. The Theorem is proved.

Theorem 11. *$F(A)$ is quasi-compact for the central topology if, and only if, A possesses the unit element.*

Proof: (a) Assume that $1 \in A$, and let $(F_\alpha)_\alpha$ be a decreasing net of compact Z-extremal subsets of $E_0(A)$, such that

$$(1) \qquad\qquad F_\alpha \cap F(A) \neq \emptyset, \quad \forall \alpha.$$

Then $F_0 \stackrel{\text{def}}{=} \cap_\alpha F_\alpha$ is a compact Z-extremal subset of $E_0(A)$, and $F_0 \cap E(A) \neq \emptyset$, because $E(A)$ is compact. Since $F_0 \cap E(A)$ is a non-empty compact Z-extremal subset of $E(A)$, from Theorem 9 we infer that $F_0 \cap F(A) \neq \emptyset$, and this shows that $(F(A); \widehat{Z}(F(A)))$ is quasi-compact.

(b) Conversely, assume that $(F(A); \widehat{Z}(F(A)))$ is quasi-compact. We shall prove that A has the unit element by adapting the proof of ([28], Proposition 3.19).

(b)$'$ A possesses a strictly positive element. Indeed, for any $a \in A^+$ let us consider the subset $F(a) = \{f \in E_0(A); f(a) = 0\}$. It is obvious that $F(a)$ is a compact face of $E_0(A)$. If A does not possess a strictly positive element, then $F(a) \cap P(A) \neq \emptyset$, for any $a \in A^+$. Since we have

$$F(a_1 + \ldots + a_n) = \overset{n}{\underset{i=1}{\cap}} F(a_i), a_1, \ldots, a_n \in A^+,$$

we infer that

$$\emptyset = \{0\} \cap F(A) = \left(\underset{a \in A^+}{\cap} F(a) \right) \cap F(A) \neq \emptyset,$$

and this is a contradiction. Assuming that $A \neq \{0\}$, let then $a_0 \in A^+, \|a_0\| = 1$, be a strictly positive element of A.

(b)$''$ For any $\alpha \in (0,1)$, let us consider the function $\phi_\alpha : [0,1] \rightarrow [0,1]$, given by

$$\phi_\alpha(t) = \begin{cases} 0, & 0 \leq t \leq \alpha \\ (1-\alpha)^{-1}(t-\alpha), & \alpha \leq t \leq 1. \end{cases}$$

Let $A_0 \subset A$ be the C^*-subalgebra generated by a_0; then we have $b \overset{\text{def}}{=} \phi_\alpha(a_0) \in A_0$ and, by the Gelfand–Naimark Representation Theorem, A_0 can be identified to the C^*-algebra $C_0(\mathfrak{M})$ of all continuous complex functions defined on the compact subset $\mathfrak{M} \subset [0,1]$, which vanish at 0 (we can always assume that $0 \in \mathfrak{M}$). For any $\alpha \in (0,1)$ we have

$$(1) \qquad F(b_\alpha) \subset \{f \in E_0(A); f(a_0) \le \alpha \|f\|\}.$$

Indeed, if $f \in E(A)$, then $f|A_0$ is a Radon probability measure μ on $A_0 \cong C_0(\mathfrak{M})$ (take into account the fact that since a_0 is strictly positive in A, the sequence $(a_0^{1/n})_{n \ge 1}$ is an approximate unit of A). If $f \in F(b_\alpha) \cap E(A)$, then

$$(2) \qquad 0 = f(b_\alpha) = \int_{\mathfrak{M}} \widetilde{b}_\alpha d\mu, \quad \alpha \in (0,1),$$

where $\widetilde{b}_\alpha$ is the Gelfand transform of b_α. From (2) we infer that $\widetilde{b}_\alpha|\operatorname{supp}\mu = 0$; i.e., we have

$$(3) \qquad \phi_\alpha(\widetilde{a}_0(m)) = 0, \quad m \in \operatorname{supp}\mu.$$

From (3) we infer that

$$\widetilde{a}_0(m) \le \alpha, \quad m \in \operatorname{supp}\mu,$$

and, therefore,

$$f(a_0) = \int_{\mathfrak{M}} \widetilde{a}_0 \, d\mu \le \alpha,$$

whence (1) immediately follows.

(b)$'''$ For any $f_0 \in F(A)$ we have $f_0 \notin F(b_{\frac{1}{2}f_0(a_0)})$. Indeed, we have $f_0(a_0) > 0$ and, therefore,

$$f_0 \in \left\{f \in E_0(A); f(a_0) > \frac{1}{2}f_0(a_0)\|f\|\right\}.$$

The assertion now immediately follows from (b)$''$.

(b)$''''$ $0 < \alpha_1 \le \alpha_2 < 1 \Rightarrow F(b_{\alpha_1}) \subset F(b_{\alpha_2})$. Indeed, we have $\phi_{\alpha_1} \ge \phi_{\alpha_2}$.

(b)$'''''$ Let us now assume that a_0 is not invertible in A. Then, for any $\alpha \in (0,1)$ there exists a $m_\alpha \in \mathfrak{M} \cap (0,\alpha)$, such that $\widetilde{a}_0(m_\alpha) = m_\alpha > 0, \widetilde{b}_\alpha(m_\alpha) = 0$. Let $p_0 \in P(A)$ be such that $p_0|A_0$ be the homomorphism $A_0 \ni a \mapsto \widetilde{a}(m_\alpha)$. We then have $p_0(b_\alpha) = 0$; i.e., $p_0 \in F(b_\alpha)$. It follows that

$$F(b_\alpha) \cap F(A) \supset F(b_\alpha) \cap P(A) \ne \emptyset, \quad \alpha \in (0,1).$$

From (b)$''''$ we now infer that

(4)
$$\left(\underset{\alpha \in (0,1)}{\cap} F(b_\alpha)\right) \cap F(A) \neq \emptyset,$$

by taking into account the assumption that $F(A)$ is quasi-compact in the central topology; but relation (4) is in contradiction to (b)$'''$. It follows that a_0 is invertible, and the Theorem is proved.

§4. Stable Projections and Liftings of Measures

The theory developed in this section is inspired by the following observation from ([26], p. 5):

Let X_1, X_2 be compact spaces and $f : X_1 \rightarrow X_2$ a continuous mapping. Let $\mu_1 \in \mathcal{M}^1_+(X_1)$ and define $\mu_2 = r_*(\mu_1) \in \mathcal{M}^1_+(X_2)$. Then the mapping

$$\mathcal{L}^1 (X_2, \mathcal{B}(X_2)) \ni \phi \mapsto \phi \circ r \in \mathcal{L}^1 (X_1, \mathcal{B}(X_1))$$

induces a *bijective isomorphism*

$$L^1(X_2, \mu_2) \ni [\phi] \mapsto [\phi \circ r] \in L^1(X_1, \mu_1)$$

if and only if, $\mu_1 \in ex\ r_*^{-1}(\{\mu_2\})$.

I. Let $(X, \sum)$ be any *measurable space* (i.e., $\sum$ is a σ-algebra of subsets of the set X). We shall denote by $\mathcal{L}(X, \sum)$ the algebra of all $\sum$-measurable complex functions defined on X, whereas $\mathcal{L}^\infty(X, \sum)$ will stand for the commutative C^*-algebra of all *bounded* functions in $\mathcal{L}(X, \sum)$, endowed with the sup-norm.

If μ is any probability measure on $\sum$, for any given $p \in [1, +\infty]$, we shall denote by $\mathcal{L}^p(X, \sum, \mu)$ the *semi-normed* complex vector space of all functions $f \in \mathcal{L}(X, \sum)$, such that $|f|^p$ is μ-integrable, if $1 \leq p < +\infty$, in which case the semi-norm is given by

$$\|f\|_p = \left(\int_X |f(x)|^p d\mu(x)\right)^{\frac{1}{p}}, \quad f \in \mathcal{L}^p \left(X, \sum, \mu\right);$$

for $p = +\infty$, we shall define

$$\mathcal{L}^\infty \left(X, \sum, \mu\right) = \mathcal{L}^\infty \left(X, \sum\right)$$

and the semi-norm will be given by

$$\|f\|_\infty = \mu - \text{vrai } \sup\{|f(x)|; x \in X\};$$

i.e., modulo the measure μ.

We shall denote by $L^p(X, \sum, \mu)$ the corresponding Banach spaces, obtained by identifying two functions $f_1, f_2 \in \mathcal{L}^p(X, \sum, \mu)$, such that $\|f_1 - f_2\|_p = 0$. In this case, we shall also write $f_1 \sim f_2(\text{mod } \mu)$.

We shall denote by $[f]$, or $C_p(f)$, the class of $f \in \mathcal{L}^p(X, \sum, \mu)$ in $L^p(X, \sum, \mu)$, thus obtaining the canonical mapping

$$C_p : \mathcal{L}^p\left(X, \sum, \mu\right) \to L^p\left(X, \sum, \mu\right), \quad p \in [1, +\infty].$$

We shall denote by $\sum(\mu)$ the *Lebesgue completion* of $\sum$ with respect to μ:

$$\sum(\mu) = \left\{ X_0 \in \mathcal{P}(X); \exists X_1, X_2 \in \sum, \text{ such that} \right.$$

$$\left. X_0 \Delta X_1 \subset X_2 \text{ and } \mu(X_2) = 0 \right\}.$$

We shall denote by $\mathcal{M}^1_+(X, \sum)$ the convex set of all probability measures $\mu : \sum \to [0, 1]$.

II. Let $\sum_1, \sum_2$ be two α-algebras of subsets of the set X, such that $\sum_1 \subset \sum_2$, and let μ_1 be a probability measure on $\sum_1$. The problem of extending μ_1 to a (probability) measure μ_2 on $\sum_2$ is very difficult, in most cases, and it depends on deep set-theoretical properties.

Assuming that such an extension exists (we could start with a given probability measure μ_2 on $\sum_2$ and consider the restriction $\mu_1 \overset{\text{def}}{=} \mu_2 | \sum_1$ of μ_2 to $\sum_1$), we obviously have the inclusions

(a) $\mathcal{L}(X, \sum_1) \subset \mathcal{L}(X, \sum_2)$,
(b) $\mathcal{L}^p(X, \sum_1, \mu_1) \subset \mathcal{L}^p(X, \sum_2, \mu_2), p \in [1, +\infty]$.

Of course, the semi-norm $\|\cdot\|_p$ on $\mathcal{L}^p(X, \sum_1, \mu_1)$ is the restriction to $\mathcal{L}^p(X, \sum_1, \mu_1)$ of the semi-norm $\|\cdot\|_p$ on $\mathcal{L}^p(X, \sum_2, \mu_2)$, for any $p \in [1, +\infty]$. We then have the commutative diagram

$$\mathcal{L}^p\left(X, \sum_1, \mu_1\right) \overset{i_p}{\longrightarrow} \mathcal{L}^p\left(X, \sum_2, \mu_2\right)$$

$$C'_p \downarrow \qquad\qquad\qquad \downarrow C''_p$$

$$L^p\left(X, \sum_1, \mu_1\right) \underset{j_p}{\longrightarrow} L^p\left(X, \sum_2, \mu_2\right)$$

135

where i_p is a proper inclusion, whereas j_p is an isometry *into* its codomain, which can be correctly and uniquely determined by the commutativity condition

$$j_p \circ C_p' = C_p'' \circ i_p, \quad p \in [1, +\infty].$$

We shall say that the extension $\mu_1 \mapsto \mu_2$ or the restriction $\mu_2 \mapsto \mu_1$, are *stable* iff j_p is *onto*; i.e., j_p is an isomorphism of Banach spaces.

Standard examples of stable extensions are:

Example 1. The Lebesgue completion of any probability space $(X, \sum, \mu)$.

Example 2. The Radon (i.e, regular Borel) extension of any Baire probability space $(X, \mathcal{B}_0(X), \mu)$, on any compact space.

We remark here that the second example is not reducible to the first; more precisely, there exist compact spaces X and Radon probability measures μ on X, such that the Lebesgue completion of the restriction of μ to $\mathcal{B}_0(X)$ does not contain $\mathcal{B}(X)$.

III. Let us now consider a measurable space $(X_1, \sum_1)$, a set X_2 and a mapping $f : X_1 \to X_2$. We can define the (full) *direct image* $r_*(\sum_1)$ of $\sum_1$ by r, given by

$$r_*\left(\sum_1\right) = \left\{ B \subset X_2; r^{-1}(B) \in \sum_1 \right\};$$

it is obvious that $r_*(\sum_1)$ is a σ-algebra of subsets of X_2. On $r_*(\sum_1)$ we can define the (full) *direct image* $r_*(\mu_1)$ of any probability measure μ_1, given on $\sum_1$, by the formula

$$r_*(\mu_1)(B) = \mu_1\left(r^{-1}(B)\right), \quad B \in r_*\left(\sum_1\right).$$

We shall say that $r_*(\mu_1)$ is the *projection* of μ_1 by r.

If $\sum_2$ is a given σ-algebra of subsets of X_2, then r is said to be $(\sum_1, \sum_2)$-*measurable* if

$$\sum_2 \subset r_*\left(\sum_1\right).$$

Remark. Given a probability measure μ_1 on $\sum_1$, and a $(\sum_1, \sum_2)$-measurable mapping $r : X_1 \to X_2$, one usually considers that $r_*(\mu_1)$ is defined on $\sum_2$ only, by restricting $r_*(\mu_1)$ to $\sum_2$, but we prefer this, more general, setting.

Now, given a probability measure μ_2 on $\sum_2$, and a $(\sum_1, \sum_2)$-measurable mapping, one could put the problem of finding a probability measure μ_1 on $\sum_1$, such that $r_*(\mu_1)| \sum_2 = \mu_2$. Such a measure μ_1 will be called a *lifting* of μ_2 by r.

Of course, the *restriction* of a probability measure to a sub-σ-algebra of sets is a particular case of a *projection*; whereas the *extension* of a probability measure, to a larger σ-algebra of sets, is a particular case of a *lifting*. This shows that, in constrast to the case of a projection, the lifting of a measure is not always possible. Therefore, the following well-known result is quite remarkable:

Let $r : X_1 \to X_2$ *be a continuous mapping between compact spaces. Then the following are equivalent:*

(a) *r is surjective;*

(b) *any Radon probability measure on X_2 can be lifted to a Radon probability measure on X_1.*

(See [26], p. 5; [28], Lemma 2.1.)

IV. Given a $(\sum_1, \sum_2)$-measurable mapping $f : X_1 \to X_2$ between two measurable spaces $(X_1, \sum_1)$ and $(X_2, \sum_2)$, and a probability measure μ_1 on $\sum_1$, let us consider the probability measure $\mu_2 = r_*(\mu_1)| \sum_2$. We can then obviously define the mapping

$$r_p : \mathcal{L}^p \left(X_2, \sum_2, \mu_2 \right) \to \mathcal{L}^p \left(X_1, \sum_1, \mu_1 \right), \quad 1 \le p \le +\infty,$$

by the formula $r_p(f) = f \circ r, f \in \mathcal{L}^p(X_2, \sum_2, \mu_2)$.

It is obvious that we can now obtain the commutative diagram

$$\mathcal{L}^p \left(X, \sum_2, \mu_2 \right) \xrightarrow{r_p} \mathcal{L}^p \left(X_1, \sum_1, \mu_1 \right)$$

$$C_p'' \downarrow \qquad\qquad\qquad \downarrow C_p'$$

$$L^p \left(X_2, \sum_2, \mu_2 \right) \xrightarrow{q_p} L^p \left(X_1, \sum_1, \mu_1 \right)$$

where q_p is uniquely determined by the commutativity condition

$$q_p \circ C_p'' = C_p' \circ r_p, \quad p \in [1, +\infty].$$

It is easy to show that q_p is an isometry *into* its codomain.

We shall say that μ_2 is a *stable projection* of μ_1 (induced by the mapping $r : X_1 \to X_2$), or, equivalently, that μ_1 is a *stable lifting* of μ_2, iff the mappings q_p are *onto* for all $p \in [1, +\infty]$.

Theorem 12.

(a) *For any $\mu_1 \in \mathcal{M}^1_+(X_1, \sum_1)$ and any $p \in [1, +\infty]$, the mapping*

$$r_p : \mathcal{L}^p \left(X_2, \sum\nolimits_2, \mu_2 \right) \ni f \mapsto f \circ r \in \mathcal{L}^p \left(X_1, \sum\nolimits_1, \mu_1 \right)$$

induces an isometry $q_p : L^p(X_2, \sum_2, \mu_2) \to L^p(X_1, \sum_1, \mu_1)$.

(b) *q_∞ is a C^*-algebra homomorphism.*

(c) *The following statements are equivalent:*

(i) *μ_1 is a stable lifting of μ_2;*

(ii) *there exists a $p \in [1, +\infty]$, such that q_p is surjective;*

(iii) *$\mu_1 \in ex \, r_*^{-1}(\{\mu_2\})$.*

(Here $r : X_1 \to X_2$ is a given $(\sum_1, \sum_2)$-measurable mapping and $\mu_2 = r_*(\mu_1)|\sum_2; r_*$ is the direct image mapping $r_* : \mathcal{M}^1_+(X_1, \sum_1) \to \mathcal{M}^1_+(X_2, \sum_2)$.

Proof: Statements (a) and (b) are obvious.

(c) It is obvious that (i) $\Rightarrow$ (ii).

(ii) $\Rightarrow$ (iii). Indeed, let $p \in [1, +\infty]$ be such that q_p be surjective. Let $\mu_1 = \frac{1}{2}(\mu_1' + \mu_1'')$ be a decomposition of μ_1, such that $\mu_1', \mu_1'' \in r_*^{-1}(\{\mu_2\})$, i.e.,

$$r_*(\mu_1') \Big| \sum\nolimits_2 = r_* \left(\mu_1'' \right) \Big| \sum\nolimits_2 = \mu_2.$$

Let $g \in \mathcal{L}^p(X_1, \sum_1, \mu_1)$ be given. Then there exists a $f \in \mathcal{L}^p(X_2, \sum_2, \mu_2)$, such that $f \circ r \sim g(\mathrm{mod}\,\mu_1)$. We then infer that $f \circ r \sim g(\mathrm{mod}\,\mu_1')$, and also, that $f \circ r \sim g(\mathrm{mod}\,\mu_1'')$. We obtain

$$\int_{X_1} g \, d\mu_1' = \int_{X_1} (f \circ r) \, d\mu_1' = \int_{X_2} f \, d\mu_2 = \int_{X_1} (f \circ r) d \, \mu_1'' = \int_{X_1} g \, d \, \mu_1'',$$

for any $g \in \mathcal{L}^p(X_1, \sum_1, \mu_1)$, and this shows that $\mu_1' = \mu_1''$; hence, we have that $\mu_1 \in ex \, r_*^{-1}(\{\mu_2\})$.

(iii) $\Rightarrow$ (i). We shall first prove that q_1 is surjective. Indeed, if

$$im \, q_1 \neq L^1(X_1, \sum\nolimits_1, \mu_1),$$

then, by the Hahn–Banach Theorem, there exists a $[g] \in (im \, q_1)^\perp, [g] \neq 0$. It is obvious that we can choose $[g]$, such that g be real, and $|g(x)| \leq 1, x \in X_1$. We then have

$$\int_{X_1} (f \circ r) g \, d\mu_1 = 0, \qquad f \in \mathcal{L}^1(X_2, \sum\nolimits_2, \mu_2),$$

and, in particular,

$$\int_{X_1} g \, d\mu_1 = 0.$$

Let us define $\mu_1' = (1+g)\mu_1, \mu_1'' = (1-g)\mu_1$. We have $\mu_1', \mu_1'' \in \mathcal{M}_+^1(X_1, \sum_1)$, and for any $f \in \mathcal{L}^1(X_2, \sum_2, \mu_2)$ we have

$$\int_{X_2} f \, d\mu_2 = \int_{X_1} (f \circ r) d\mu_1 = \int_{X_1} (f \circ r)(1+g) d\mu_1 = \int_{X_1} (f \circ r) d\mu_1',$$

$$\int_{X_2} f \, d\mu_2 = \int_{X_1} (f \circ r) d\mu_1 = \int_{X_1} (f \circ r)(1-g) d\mu_1 = \int_{X_1} (f \circ r) d\mu_1'',$$

and this shows that $r_*(\mu_1') = \mu_2 = r_*(\mu_1'')$. It follows that $\mu_1', \mu_1'' \in r_*^{-1}(\{\mu_2\})$, and since we have $\mu_1 = \frac{1}{2}(\mu_1' + \mu_1'')$, we infer that $\mu_1' = \mu_1 = \mu_1''$. We infer that $[g] = 0$ in $L^\infty(X_1, \sum_1, \mu_1)$, and this is a contradiction.

Let us now remark that the mapping q_∞ is (w^*, w^*)-continuous. Indeed, for any $f \in \mathcal{L}^1(X_2, \sum_2, \mu_2)$ and any $g \in \mathcal{L}^\infty(X_2, \sum_2, \mu_2)$ we have

$$\int_{X_2} fg \, d\mu_2 = \int_{X_1} (f \circ r)(g \circ r) d\mu_1$$

and, since the mapping q_1 is surjective, we infer that if $[g] \to 0$ in

$$\sigma \left(L^\infty \left(X_2, \sum_2, \mu_2 \right); L^1 \left(X_2, \sum_2, \mu_2 \right) \right),$$

then $q_\infty([g]) \to 0$ in the topology

$$\sigma \left(L^\infty \left(X_1, \sum_1, \mu_1 \right); L^1 \left(X_1, \sum_1', \mu_1 \right) \right).$$

Since q_∞ is an isometry, it follows that the closed unit ball in $im \, q_\infty$ is the image by q_∞ of the closed unit ball in $L^\infty(X_2, \sum_2, \mu_2)$; hence, it is w^*-compact, by the Alaoglu–Bourbaki Theorem.

From the Krein–Šmulian Theorem we now infer that $im \, q_\infty$ is w^*-closed in

$$L^\infty \left(X_1, \sum_1, \mu_1 \right).$$

If $im \, q_\infty \neq L^\infty(X_1, \sum_1, \mu_1)$, then we could find an $f_1 \in \mathcal{L}^1(X_1, \sum_1, \mu_1)$, such that $[f_1] \neq 0$, and

$$\int_{X_1} f_1(g \circ r) \, d\mu_1 = 0, \qquad g \in \mathcal{L}^\infty(X_2, \sum_2, \mu_2).$$

139

Since q_1 is surjective, we infer that there exists an $f_2 \in \mathcal{L}^1(L_2, \sum_2, \mu_2)$ such that $q_1([f_2]) = [f_1]$, and, therefore we had

$$0 = \int_{X_1} f_1(g \circ r)\, d\mu_1 = \int_{X_1} (f_2 \circ r)(g \circ r) d\mu_1 = \int_{X_2} f_2 g \, d\mu_2,$$

for any $g \in \mathcal{L}^\infty(X_2 \sum_2, \mu_2)$. This implies that $[f_2] = 0$ and, therefore, $[f_1] = 0$, contrary to the assumption.

We have thus proved that q_∞ is surjective. Let us now remark that for any $A_1 \in \sum_1$ there exists an $A_2 \in \sum_2$, such that $\chi_{A_1} \sim \chi_{r^{-1}(A_2)} (\bmod \ \mu_1)$. Indeed, since q_∞ is surjective, there exists a $g \in \mathcal{L}^\infty(X_2, \sum_2, \mu_2)$, such that

$$0 = \int_{X_1} \left| \chi_{A_1} - g \circ r \right| d\mu = \int_{A_1} |1 - g \circ r| \, d\mu_1 + \int_{\mathcal{C}A_1} (|g| \circ r) \, d\mu_1.$$

Let $A_1' = \{x \in A_1; (g \circ r)(x) = 1\}$ and $A_1'' = \{x \in \mathcal{C}A_1; (g \circ r)(x) = 0\}$. We then have $\mu_1(A_1) = \mu_1(A_1')$ and $\mu_1(\mathcal{C}A_1) = \mu_1(A_1'')$. Let $B_1 = \{y \in X_2; g(y) = 1\}$ and $B_0 = \{y \in X_2; g(y) = 0\}$. We have

$$A_1' = r^{-1}(B_1) \cap A_1 \text{ and } A_1'' = r^{-1}(B_0) \cap \mathcal{C}A_1.$$

We infer that we have

$$\int_{X_1} \left| \chi_{r^{-1}(B_1)} - \chi_{A_1} \right| d\mu_1 = \int_{A_1} \left| \chi_{r^{-1}(B_1)} - 1 \right| d\mu_1 + \int_{\mathcal{C}A_1} \chi_{r^{-1}(B_1)} d\mu_1$$

$$= \int_{A_1} |\chi_{A'} - 1| \, d\mu_1 + \int_{A_1''} \chi_{r^{-1}(B_1)} d\mu_1 = 0,$$

and this shows that $\chi_A \sim \chi_{r^{-1}(B_1)} (\bmod \ \mu_1)$.

Let us now prove that q_p is surjective, for any $p \in [1, +\infty)$. Indeed, if

$$f \in \mathcal{L}^p \left(X_1; \sum_1, \mu_1 \right),$$

then there exists a sequence $(A_n)_{n \geq 0}$ of sets $A_i \in \sum_i; i \in \mathbb{N}$, such that

(σ) $A_i \cap A_j = \emptyset$, for $i \neq j$;

(β) $\cap_{i \geq 0} A_i = X_1$;

(γ) $f \chi_{A_i} \in \mathcal{L}^\infty(X_1, \sum_1, \mu_1)$, $\quad i \geq 0$.

We infer that for any $i \in \mathbb{N}$ there exists a $g_i \in \mathcal{L}^\infty(X_2, \sum_2, \mu_2)$, such that $g_i \circ r \sim f \chi_{A_i}$. It is easy to prove that the series $\sum_{i=0} g_i$ is norm converging in $\mathcal{L}^p(X_2, \sum_2, \mu_2)$. If we define $g = \sum_{i=0}^\infty g_i$, we have $g \in \mathcal{L}^p(X_2, \sum_2, \mu_2)$, and $r_p([g]) = [f]$. The Theorem is proved.

In the same setting as above, we can also prove the following criterion of stability.

Theorem 13. *The following statements are equivalent:*

(a) μ_1 *is a stable lifting of* μ_2;

(b) *for any* $A_1 \in \sum_1$ *there exists an* $A_2 \in \sum_2$, *such that* $\mu_1(A_1 \Delta r^{-1}(A_2)) = 0$.

Proof: The implication (a) $\Rightarrow$ (b) follows from the fact that the stability of μ_1 with respect to r implies that q_∞ is surjective, whereas (b) $\Rightarrow$ (a) follows from the fact that (b) implies the surjectivity of q_∞, as one could easily see by approximating any function in $\mathcal{L}^\infty(X_1, \sum_1, \mu_1)$ by simple functions, uniformly.

V. The well-known method of introducing measures on possibly non-measurable subsets gives another instance of a stable lifting.

Example 3. Let $(X_2, \sum_2, \mu_2)$ be any probability space and consider any subset $X_1 \subset X_2$, such that $\mu_2^*(X_1) = 1$. Define

$$\sum_1 = \left\{ M \cap X_1 ; M \in \sum_2 \right\};$$

obviously, $\sum_1$ is a σ-algebra of subsets of X_1.

Since $\mu_2^*(M \cap X_1) = \mu_2(M), M \in \sum_2$, we can define a probability measure $\mu_1 : \sum_1 \to [0,1]$ by the formula

$$\mu_1(M \cap X_1) = \mu_2(M), \quad M \in \sum_2.$$

If $r : X_1 \to S_2$ is the inclusion mapping, then it is easy to see that μ_1 is a stable lifting of μ_2, by r.

Consider now a stable extension $(X_2, \sum_3, \mu_3)$ of $(X_2, \sum_2, \mu_2)$, where $\sum_2 \subset \sum_3$ and $\mu_3 | \sum_2 = \mu_2$. We can define correctly a canonical bijective isometry $r_p : L^p(X_2, \sum_3, \mu_3) \to L^p(X_1, \sum_1, \mu_1)$, for any $p \in [1, +\infty]$, in the following manner.

For any class $[f] \in L^p(X_2, \sum_3, \mu_3)$ choose a representative $f \in \mathcal{L}^p(X_2, \sum_2, \mu_2)$ and restrict it to X_1; then $[f|X_1]$ is the correctly defined image $r'_p([f])$ of $[f]$, and the mapping r'_p is easily seen to be a bijective isometry.

This setting is often encountered in Choquet Theory and, as an application, in Reduction Theory. Namely, let X_2 be any compact convex set in a Hausdorff locally convex topological real vector space, and let $X_1 = ex\ X_2$ be its extremal boundary. Let μ_3 be any Choquet maximal Radon probability measure, defined on $\sum_3 = \mathcal{B}(X_2)$, and let $\sum_2 = \mathcal{B}_0(X_2), \mu_2 = \mu_3 | \sum_2$. Then $\mu_2^*(X_1) = 1$, by the Choquet–Bishop–de Leeuw Theorem and the preceding considerations can be applied.

Below we shall encounter another instance of this example, by considering the set of the factorial states of a C^*-algebra, and central measures.

Remark. The indiscriminate restriction mapping

$$\mathcal{L}^p(X_2, \textstyle\sum_3, \mu_3) \ni f \mapsto f|X_1$$

is not legitimate, in general, for defining the mapping r'_p.

The following Theorem shows that there exist maximal stable extensions for any probability measure.

Theorem 14. *For any probability space* (X_0, Σ_0, μ_0) *there exist maximal stable extensions* (X_0, Σ_1, μ_1).

Proof: Let $\mathcal{S}$ be a totally ordered set of stable extensions $(X_0, \Sigma, \mu_\Sigma)$ of (X_0, Σ_0, μ_0), such that for any $(X_0, \Sigma', \mu_{\Sigma'},), (X_0, \Sigma'', \mu_{\Sigma''}) \in \mathcal{S}$, either $\Sigma' \subset \Sigma''$ and $\mu_{\Sigma''}|\Sigma' = \mu_{\Sigma'}$, or $\Sigma'' \subset \Sigma'$ and $\mu_{\Sigma'}|\Sigma'' = \mu_{\Sigma''}$.

Of course, we can assume that $(X_0, \Sigma_0, \mu_0) \in \mathcal{S}$ and it is easy to prove that if $(X_0, \Sigma', \mu_{\Sigma'}), (X_0, \Sigma'', \mu_{\Sigma''}) \in \mathcal{S}$, and $\Sigma' \subset \Sigma''$, then $(X_0, \Sigma'', \mu_{\Sigma''})$ is a stable extension of $(X_0, \Sigma', \mu_{\Sigma'})$.

We can define on the algebra $\widetilde{\Sigma} = \cup\{\Sigma; \Sigma \in \mathcal{S}\}$ the finitely additive set function $\widetilde{\mu} : \widetilde{\Sigma} \to [0,1]$ by

$$\widetilde{\mu}(M) = \mu_\Sigma(M), \text{ if } M \in \Sigma,$$

where $(X_0, \Sigma, \mu_\Sigma)$ is a suitably chosen extension in $\mathcal{S}$ (for any $M \in \widetilde{\Sigma}$ there exists such an extension).

Let us now consider the outer measure $\widetilde{\mu}^*$ corresponding to $\widetilde{\mu}$; i.e., we define

$$\widetilde{\mu}^*(M) = \inf\left\{ \sum_{i=0}^{\infty} \widetilde{\mu}(M_i); M_i \in \widetilde{\Sigma}, \cup_{i=0}^{\infty} M_i \supset M \right\},$$

for any $M \subset X_0$. Let $\Sigma_1 = \{M \in \mathcal{P}(X_0); \exists M_0 \in \Sigma_0, \text{ such that } \widetilde{\mu}^*(M \Delta M_0) = 0\}$, and define $\mu_1 = \mu^*|\Sigma_1$. Then (X_0, Σ_1, μ_1) is a probability space, and it is a stable extension of any $(X_0, \Sigma, \mu_\Sigma) \in \mathcal{S}$. The application of Zorn's Lemma finishes the proof.

§5. Stable Liftings of Orthogonal Measures

In order to prove the topological properties of the central measures we have in view, we need to establish the possiblity of lifting orthogonal measures as orthogonal measures.

I. Let $\pi : A \to B$ be a homomorphism of C^*-algebras. We do not assume them to possess the unit element; whereas, if they possess it, we do not assume that $\pi(1) = 1$.

Let $\pi^* : B^* \to A^*$ be the adjoint mapping and denote by $s : E_0(B) \to E_0(A)$ its restriction-corestriction to $E_0(B)$ and $E_0(A)$.

Theorem 15.

(a) $s(E_0(B))$ is a compact face of $E_0(A)$.

(b) $s(E(B))$ is a face of $E_0(A)$.

(c) If $1 \in B$, then $s(E(B))$ is a compact face of $E_0(A)$.

(d) If $1 \in A, 1 \in B$ and $\pi(1) = 1$, then $s(E(B))$ is a compact face of $E(A)$.

Proof: (a)   Since the mapping s is affine and w^*-continuous, it is obvious that $s(E_0(B))$ is a compact convex subset of $E_0(A)$. Let $f \in s(E_0(B))$ and let $f = tf' + (1-t)f'', 0 < t < 1, f', f'' \in E_0(A)$ be a decomposition of f. Then we have $f' | \ker \pi_f = f'' | \ker \pi_f = 0$ and, therefore, there exist $g_0', g_0'' \in \pi(A)^*$ such that $f' = g_0' \circ \pi, f'' = g_0'' \circ \pi$. From $\pi(A)^+ = \pi(A^+)$ we infer that $g_0' \geq 0, g_0'' \geq 0$; whereas from $\pi(A)_1^+ = \pi(A_1^+)$ we infer that $\|f'\| = \|g_0'\|, \|f''\| = \|g_0''\|$. If we extend g_0' to $g' \in E_0(B)$, and g_0'' to $g'' \in E_0(B)$, we have $s(g') = f' \in s(E_0(B))$ and $s(g'') = f'' \in s(E_0(B))$.

(b)   If $f \in s(E(B))$, with the same notations as above, since $\|g_0'\| \leq 1$ and $\|g_0''\| \leq 1$, we can find extensions g' of g_0', and g'' of g_0'', such that $\|g'\| = \|g''\| = 1$; then we have $s(g') = f' \in s(E(B))$ and $s(g'') = f'' \in s(E(B))$.

(c)   If $1 \in B$, then $E(B)$ is a compact subset of $E_0(B)$. It folllows that $s(E(B))$ is a compact subset of $E_0(A)$, and a face of $E_0(A)$, by (b).

(d)   Under the assumptions, we have $s(E(B)) \subset E(A)$.

The Theorem is proved.

Theorem 16.

(a) For any orthogonal measure $\mu \in \mathcal{M}_+^1(E_0(A))$ such that $b(\mu) \in s(E_0(B)) \cap E(A)$, there exists an orthogonal measure $\nu \in \mathcal{M}_+^1(E_0(B))$ such that $s_*(\nu) = \mu$ and ν be a stable lifting of μ.

(b) If μ is maximal orthogonal, then ν can be chosen to be, moreover, maximal orthogonal.

Proof:

(a) Since, by Theorem 15, $s(E_0(B))$ is a compact face of $E_0(A)$, we have $\mu(s(E_0(B))) = 1$ (see [28], Proposition 1.5). We then infer that $s_*^{-1}(\{\mu\})$ is a non-empty compact convex subset of $\mathcal{M}^1_+(E_0(B))$. Let $\nu \in ex\ s_*^{-1}(\{\mu\})$. By Theorem 12 we infer that ν is a stable lifting of μ. We shall prove that ν is also orthogonal.

(a$'$) If $[\phi] \in L^\infty(E_0(A), \mathcal{B}(E_0(A)), \mu)$ is a projection, then

$$q_\infty([\phi]) \in L^\infty(E_0(B), \mathcal{B}(E_0(B)), \nu)$$

is also a projection, and any projection in $L^\infty(E_0(B), \mathcal{B}(E_0(B)), \nu)$ is of this form, on account of the fact that ν is a stable lifting of μ.

Since μ is orthogonal, $P = K_\mu([\phi])$ is a projection in $\pi_{b(\mu)}(A)'$ (see [28], Theorem 3.1).

From the fact that $s_*(\nu) = \mu$, we infer that $s(b_B(\nu)) = b_A(\mu)$ and this implies that $\|b_B(\nu)\| = 1$, since $1 = \|b_A(\mu)\| \leq \|b_B(\nu)\| \leq 1$.

By ([12], 2.4.9) we infer that $\pi_{b(\mu)}$ can be identified with $\pi_{b(\nu)} \circ \pi : A \to \mathcal{L}(e_0 H_{b(\nu)})$, where e_0 is the projection onto the subspace $\overline{\pi_{b(\nu)}(\pi(A))\xi_{b(\nu)}}$, whereas $\xi^0_{b(\mu)}$ can be iden- tified with $\xi^0_{b(\nu)}$. By ([28], Lemma 3.1), $Q = K_\nu(q_\infty([\phi]))$ is an operator in $\pi_{b(\nu)}(\pi(A))'$ such that $0 \leq Q \leq 1$.

(a$''$) Let us now prove that

$$(1) \qquad\qquad K_\mu([\phi]) = e_0 K_\nu(q_\infty([\phi]))e_0,$$

for any $[\phi] \in L^\infty(E_0(A), \mathcal{B}(E_0(A)), \mu)$. Indeed, we have

$$\left(K_\mu\left([\phi]\right)\pi_{b(\mu)}(a)\xi^0_{b(\mu)} \;\middle|\; \xi^0_{b(\mu)} \right) = \int_{E_0(A)} \phi\lambda_A(a)\ d\mu$$
$$= \int_{E_0(B)} (\phi \circ s)(\lambda_A(a) \circ s)d\nu$$
$$= \int_{E_0(B)} (\phi \circ s)\lambda_B(\pi(a))\ d\nu$$
$$= \left(K_\nu([\phi \circ s])\pi_{b(\nu)}(\pi(a))\xi^0_{b(\nu)} \;\middle|\; \xi^0_{b(\nu)} \right), \quad a \in A,$$

and this shows that

$$K_\mu([\phi]) = e_0 K_\nu([\phi \circ s])e_0,$$

as requested.

(a$'''$) From formula (1) we infer that $P = e_0 Q e_0$ and, therefore, we have

$$e_0 Q e_0 Q e_0 = e_0 Q e_0, Q e_0 Q \leq Q^2 \leq Q.$$

144

We infer that

$$e_0(Q - Qe_0Q)e_0 = 0,$$

and, therefore, we have that

$$(Q - Qe_0Q)e_0 = 0,$$

whence we get that

$$Qe_0 = Qe_0Qe_0, e_0Q = e_0Qe_0Q.$$

From $Q^2 \leq Q$ we infer that $e_0Q^2e_0 \leq e_0Qe_0$ and, if we denote

$$R = (Qe_0 - e_0Q)^* (Qe_0 - e_0Q),$$

we have that

$$0 \leq R = e_0Q^2e_0 - e_0Q - Qe_0 + Qe_0Q.$$

We infer that

$$0 \leq e_0Re_0 = e_0Q^2e_0 - e_0Qe_0 - e_0Qe_0 + e_0Qe_0Qe_0$$
$$= e_0Q^2e_0 - e_0Qe_0 \leq 0$$

and this shows that $e_0Re_0 = 0$. We infer that $Re_0 = 0$ and, therefore,

$$(Qe_0 - e_0Q)e_0 = 0,$$

whence we get that

$$Qe_0 = e_0Qe_0;$$

this implies that

(2) $$Qe_0 = e_0Q.$$

From (2) we infer that $Q^2e_0 = Qe_0$ and, since $e_0\xi^0_{b(\nu)} = \xi^0_{b(\nu)}$, we get that

(3) $$Q^2\xi^0_{b(\nu)} = Q\xi^0_{b(\nu)}.$$

Since $Q \in \pi_{b(\nu)}(B)'$, and since $\xi^0_{b(\nu)}$ is cyclic for $\pi_{b(\nu)}(B)$, from (3) we infer that $Q^2 = Q$; i.e., Q is a projection.

From ([28], Theorem 3.1) we now infer that ν is orthogonal. (We have also to take into consideration the fact that ν is a stable lifting of μ.)

(b) Let us now assume, moreover, that μ is maximal orthogonal on $E_0(A)$. Since the set $b_B(s_*^{-1}(\{\mu\})) \subset E_0(B)$ is compact, convex, and non-empty, we can find a $g_0 \in ex\ b_B(s_*^{-1}(\{\mu\}))$. Let us define

$$M(g_0) = s_*^{-1}(\{\mu\}) \cap b_B^{-1}(\{g_0\}):$$

It is obvious that $M(g_0)$ is a non-empty compact face of $s_*^{-1}(\{\mu\})$, and, therefore, we have

$$\emptyset \neq ex\ M(g_0) = M(g_0) \cap (ex\ s_*^{-1}(\{\mu\})).$$

Let us choose $\nu \in ex\ M(g_0)$. By part (a) of the Theorem, ν is an orthogonal measure on $E_0(B)$.

If $\nu_1 \in \mathcal{M}_+^1(E_0(B))$ is a maximal orthogonal probability measure on $E_0(B)$, such that

$$(4) \qquad\qquad\qquad\qquad \nu \prec \nu_1,$$

then $\nu \sim \nu_1$ and, therefore, $b_B(\nu) = b_B(\nu_1) = g_0$. It follows that $\nu_1 \in b_B^{-1}(\{g_0\})$. On the other hand, from (4) we infer that

$$(5) \qquad\qquad\qquad\qquad \mu = s_*(\nu) \prec s_*(\nu_1),$$

and, since μ is maximal orthogonal on $E_0(A)$, from (5) it follows that $\mu = s_*(\nu_1)$; i.e., $\nu_1 \in s_*^{-1}(\{\mu\})$. We infer that $\nu_1 \in M(g_0)$. Since ν_1 is orthogonal, from ([28], Lemma 3.3 and Corollary 1 to Theorem 3.1), we infer that ν_1 is simplicial; i.e.,

$$\nu_1 \in ex\ \mathcal{M}_+^1(E_0(B); g_0).$$

We then infer that $\nu_1 \in ex\ M(g_0)$, and this implies that

$$\nu_1 \in ex(s_*^{-1}(\{\mu\}));$$

hence, ν_1 is a stable lifting of μ. The Theorem is proved.

Remark. If we consider the setting of the proof of the preceding theorem, since finite linear combinations of projections in the C^*-algebra $L^\infty(E_0(A), \mathcal{B}(E_0(A)), \mu)$ are uniformly dense in this space, from formulas (1) and (2) we infer that $e_0 \in \mathcal{C}_\nu'$ and $\mathcal{C}_\mu = \mathcal{C}_\nu e_0$. Moreover, the mapping $\mathcal{C}_\nu \ni c \mapsto ce_0 \in \mathcal{C}_\mu$ is an isomorphism between

146

the abelian von Neumann algebras $\mathcal{C}_\nu$ and $\mathcal{C}_\mu$. We also remark that Theorem 16 partly contains a result of Anderson and Bunce (see [2], Theorem 5).

II. We want to mention here two instances of the situation described in the preceding theorem.

(a) Consider an arbitrary C^*-algebra A and let $f_0 \in E(A)$ be given. We can consider the representation $\pi_f : A \to \mathcal{L}(H_{f_0})$. Let $\mathcal{C} \subset \pi_{f_0}(A)'$ be any maximal abelian von Neumann subalgebra, and define B_0 to be the C^*-algebra $C^*(\pi_{f_0}(A), \mathcal{C})$, generated by $\pi_{f_0}(A)$ and $\mathcal{C}$. We have then the C^*-homomorphism $\pi : A \to B_0$ obtained by corestricting π_{f_0} to B. We then have that

$$B_0' = \pi_{f_0}(A)' \cap \mathcal{C}' = \mathcal{C},$$

and we can consider the mapping $s : E_0(B) \to E_0(A)$, defined as in §5.1. If we denote $g_0 = \omega_{\xi_{f_0}^0}|B$, then $s(g_0) = f_0$. (For any $\xi \in H$ we denote by ω_ξ the positive linear functional on $\mathcal{L}(H)$ given by $\omega_\xi(x) = (x\xi|\xi), x \in \mathcal{L}(H)$.)

Let ν be the central measure on $E_0(B)$, corresponding to g_0. It is the *greatest* orthogonal measure on $E_0(B)$ whose barycenter is g_0. Of course, $E(B)$ is compact, and $\nu(E(B)) = 1$.

On the other hand, $\mu = s_*(\nu)$ is the *maximal* orthogonal measure on $E_0(A)$, whose barycenter is f_0, and which corresponds to $\mathcal{C}$ (see [28], Lemma 3.8). By the Corollary to this Lemma, ν is a stable lifting of μ.

(b) Consider now an arbitrary C^*-algebra A. Let $f_0 \in E(A)$ be given and define B_{f_0} to be the C^*-algebra $C^*(\pi_{f_0}(A), \pi_{f_0}(A)')$ generated by $\pi_{f_0}(A)$ and $\pi_{f_0}(A)'$. We have

$$B_{f_0}' = \pi_{f_0}(A)' \cap \pi_{f_0}(A)'' \subset B_{f_0}.$$

Hence, B_{f_0}' is abelian, and it is the center of B_{f_0}.

If $g_0 = \omega_{\xi_{f_0}^0}|B_{f_0}$, then there exists the *greatet* orthogonal probability measure ν_{g_0} on $E_0(B_{f_0})$ whose barycenter is at g_0, and which corresponds to the *greatest* abelian von Neumann subalgebra B_{f_0}' of B_{f_0}'.

We can also consider the C^*-homomorphism $\pi : A \to B_{f_0}$ obtained by corestricting π_{f_0}, and the corresponding mapping $s : E_0(B_{f_0}) \to E_0(A)$, given by $s(g) = g \circ \pi, g \in E_0(B_{f_0})$. Of course, $E(B_{f_0})$ is compact and $\nu_{g_0}(E(B_{f_0})) = 1$.

The following theorem extends to the general case of C^*-algebras A, possibly not possessing the unit element, a well-known theorem (see [24], Proof of Theorem 3.1.8).

Theorem 17. $s_*(\nu_{g_0}) = \mu_{f_0}$; i.e., $s_*(\nu_{g_0})$ is the central measure on $E_0(A)$, whose barycenter is at f_0.

Proof: (a) We shall identify π_{g_0} with the identical representation of B_{f_0} and we can assume that $\xi^0_{g_0} = \xi^0_{f_0}$. It is clear that $b(s_*(\nu_{g_0})) = s(b(\nu_{g_0})) = s(g_0) = f_0$.
(b) From ([28], Lemma 3.3) we infer that

(1) $$K_{s_*(\nu_{g_0})}(\|\cdot\|) = 1.$$

(c) For any $\phi \in \mathcal{L}^\infty(E_0(A), \mathcal{B}(E_0(A)), s_*(\nu_{g_0}))$ we have that

$$\left(K_{s_*(\nu_{g_0})}([\phi])\pi_{f_0}(a)\xi^0_{f_0}\,|\,\xi^0_{f_0}\right) = \int_{E_0(A)} \phi\lambda_A(a)\ ds_*(\nu_{g_0})$$

$$= \int_{E_0(B_{f_0})} (\phi\circ s)(\lambda_A(a)\circ s)\ d\nu_{g_0}$$

$$= \int_{E_0(B_{f_0})} (\phi\circ s)\lambda_B\left(\pi_{f_0}(a)\right)\ d\nu_{g_0}$$

$$= \left(K_{\nu_{g_0}}([\phi\circ s])\pi_{f_0}(a)\xi^0_{f_0}\,|\,\xi^0_{f_0}\right), \quad a \in A,$$

and since $\xi^0_{f_0}$ is cyclic for $\pi_{f_0}(A)$, we infer that

(2) $$K_{s_*(\nu_{g_0})}([\phi]) = K_{\nu_{g_0}}(q_\infty([\phi])),$$

for any $\phi \in \mathcal{L}^\infty(E_0(A), \mathcal{B}(E_0(A)), s_*(\nu_{g_0}))$. From formula (2) we immediately infer that the measure $s_*(\nu_{g_0}) \in \mathcal{M}^1_+(E_0(A); f_0)$ is orthogonal.

Let e be the orthogonal projection on $\overline{B'_{f_0}\xi^0_{f_0}} \subset H_{f_0}$. From formula (1) and from ([28], Lemma 3.6), we infer that

$$\int_{E_0(A)} \lambda_A(a_1)\lambda_A(a_2)\dots\lambda_A(a_n)\ ds_*(\nu_{g_0})$$

$$= \int_{E_0(B_{f_0})} \lambda_{B_{f_0}}\left(\pi_{f_0}(a_1)\right)\dots\lambda_{B_{f_0}}\left(\pi_{f_0}(a_n)\right)\ d\nu_{g_0}$$

$$= \left(K_{\nu_{g_0}}\left(\lambda_B\left(\pi_{f_0}(a_1)\right)\right)\dots K_{\nu_{g_0}}\left(\lambda_B\left(\pi_{f_0}(a_n)\right)\right)\xi^0_{f_0}\,|\,\xi^0_{f_0}\right)$$

$$= \left(e\pi_{f_0}(a_1)e\pi_{f_0}(a_2)e\dots e\pi_{f_0}(a_n)e\xi^0_{f_0}\,|\,\xi^0_{f_0}\right)$$

$$= \left(K_{\nu_{f_0}}\left(\lambda_A(a_1)\lambda_A(a_2)\dots\lambda_A(a_n)\right)\xi^0_{f_0}\,|\,\xi^0_{f_0}\right)$$

$$= \int_{E_0(A)} \lambda_A(a_1)\lambda_A(a_2)\dots\lambda_A(a_n)\ d\mu_{f_0}, \quad a_1, a_2, \dots, a_n \in A.$$

From the Stone–Weierstrass Theorem, by taking into account the fact that $s_*(\nu_{g_0})$ and μ_{f_0} are both supported by $E(A)$, we infer that $s_*(\nu_{g_0}) = \mu_{f_0}$. The Theorem is proved.

148

Theorem 18. ν_{g_0} *is a stable orthogonal lifting of* μ_{f_0}.

Proof: From the preceding Theorem and from formula (2) in its proof we infer that

$$(3) \qquad\qquad K_{\mu_{f_0}}([\phi]) = K_{\nu_{g_0}}(q_\infty([\phi])),$$

for any $\phi \in \mathcal{L}^\infty(E_0(A), \mathcal{B}(E_0(A)), \mu_{f_0})$. Let $\psi \in \mathcal{L}^\infty(E_0(B_{f_0}), \mathcal{B}(E_0(B_{f_0})), \nu_{g_0})$ then $K_{\nu_{g_0}}([\psi]) \in B'_{f_0}$ and, therefore, there exists a function $\phi \in \mathcal{L}^\infty(E_0(A), \mathcal{B}(E_0(A)), \mu_f)$, such that $K_{\mu_{f_0}}([\phi]) = K_{\nu_{g_0}}([\psi])$. Formula (3) now implies that $K_{\nu_{g_0}}([\psi]) = K_{\nu_{g_0}}(q_\infty([\phi]))$ and, since $K_{\nu_{g_0}}$ is injective, we infer that $q_\infty([\phi]) = [\psi]$. Q.E.D.

III. We return to the setting of Theorem 16; i.e., $\pi : A \to B$ is any homomorphism of C^*-algebras. Let $g \in E(B)$ and let $\nu \in \mathcal{M}_+^1((E_0(B))$ be any orthogonal measure, such that $b(\nu) = g$. We have $\overline{\pi_g(\pi(A))\xi_g^0} = e_0 H_g$ where $e_0 \in \pi_g(\pi(A))'$ is a projection.

Theorem 19. *Let* $\mu = s_*(\nu)$ *and assume that* $s(g) \in E(A)$. *Then the following statements are equivalent:*
(a) $e_0 \in \mathcal{C}'_\nu$;
(b) *the measure* μ *is orthogonal and* ν *is a stable lifting of* μ.

Proof: Let $f = s(g)$; then the representation $\pi_f : A \to \mathcal{L}(H_f)$ can be identified with

$$A \ni a \mapsto \pi_g(\pi(a))e_0 \in \mathcal{L}(e_0 H_g),$$

whereas $\xi_{f_0}^0$ can be identified with ξ_g^0, since $\|f\| = 1$ implies that $e_0 \xi_g^0 = \xi_g^0$.

By formula (1) in the proof of Theorem 16 we have

$$K_\mu([\phi]) = e_0 K_\nu(q_\infty([\phi]))e_0, [\phi] \in L^\infty(E_0(A), \mathcal{B}(E_0(A)), \mu).$$

(a) $\Rightarrow$ (b). Indeed, if $e_0 \in \mathcal{C}'$, then from formula (1) we immediately infer that the mapping K_μ is multiplicative; hence, μ is an orthogonal measure on $E_0(A)$.

From the fact that $f \in E(A)$, we infer that

$$K_\mu(\|\cdot\|) = 1.$$

From ([28], Lemma 3.6) we infer that

$$\int_{E_0(A)} \lambda_A(a_1)\lambda_A(a_2)\dots\lambda_A(a_n) \, d\mu = \int_{E_0(B)} \lambda_B(\pi(a_1))\lambda_B(\pi(a_2))\dots\lambda_B(\pi(a_n)) \, d\nu$$
$$= \left(K_\nu([\lambda_B(\pi(a_1))])K_\nu([\lambda_B(\pi(a_2))])\dots K_\nu([\lambda_B(\pi(a_n))])\xi_g^0|\xi_g^0\right)$$
$$= \left(e_\nu \pi_g(\pi(a_1))e_\nu\pi_g(\pi(a_2))e_\nu\dots e_\nu\pi_g(\pi(a_n))e_\nu\xi_g^0|\xi_g^0\right), \ a_1, a_2, \dots, a_n \in A,$$

149

where e_ν is the projection onto $\overline{\mathcal{C}_\nu \xi_g^0} \subset H_g$.

On the other hand, from $\mathcal{C}_\nu \subset \pi_g(B)'$ and from $\pi_g(\pi(A)) \subset \pi_g(B)$, we infer that $\pi_g(B)' \subset \pi_g(\pi(A))'$ and, therefore, we have

$$\mathcal{C}_\mu \subset \mathcal{C}_\nu e_0 \subset e_0 \pi_g(\pi(A))' e_0.$$

Since $\mathcal{C}_\nu e_0$ is an abelian von Neumann subalgebra in $e_0 \pi_g(\pi(A))' e_0 = \pi_f(A)' \subset \mathcal{L}(e_0 H_g)$, from ([28], Theorem 3.3) we infer that there exists an orthogonal measure $\mu \in \mathcal{M}_+^1(E_0 (A))$, such that $\mu \prec \mu_0$ and $\mathcal{C}_{\mu_0} = \mathcal{C}_\nu e_0$. From $e_0 \in \mathcal{C}_\nu'$ we infer that $e_\nu \leq e_0$, whereas from $e_0 \xi_g^0 = \xi_g^0$ we infer that

$$\overline{\mathcal{C}_{\mu_0} \xi_g^0} = \overline{\mathcal{C}_\nu e_0 \xi_g^0} = \overline{\mathcal{C}_\nu \xi_g^0} = e_\nu H_g;$$

hence, $e_{\mu_0} = e_\nu$. Since $\|b(\mu_0)\| = 1$, from ([28], Lemma 3.6) we infer that

$$(3) \qquad \int_{E_0(A)} \lambda_A(a_1)\lambda_A(a_2)\ldots\lambda_A(a_n) \, d\mu_0 = \left(e_{\mu_0}\pi_f(a_1)e_{\mu_0}\ldots e_{\mu_0}\pi_f(a_n)e_{\mu_0}\xi_f^0|\xi_f^0\right)$$
$$= \left(e_\nu\pi_g\left(\pi(a_1)\right)e_\nu \ldots e_\nu\pi_g\left(\pi(a_n)\right)e_\nu\xi_g^0|\xi_g^0\right),$$
$$a_1, a_2, \ldots, a_n \in A.$$

From (2) and (3) we infer that

$$(4) \qquad \int_{E_0(A)} \lambda_A(a_1)\lambda_A(a_2)\ldots\lambda_A(a_n) \, d\mu = \int_{E_0(A)} \lambda_A(a_1)\lambda_A(a_2)\ldots\lambda_A(a_n) \, d\mu_0,$$

for any $a_1, a_2, \ldots, a_n \in A$. Since $\mu(E(A)) = \mu_0(E(A)) = 1$, from (4), with the help of the Stone–Weierstrass Theorem, we infer that $\mu = \mu_0$; hence, we have that $\mathcal{C}_\mu = \mathcal{C}_\nu e_0$.

Let us now consider the commutative diagram

$$\begin{array}{ccc}
L^\infty(E_0(B), \mathcal{B}(E_0(B)), \nu) & \xrightarrow{K_\nu} & \mathcal{C}_\nu \\
q_\infty \uparrow & & \downarrow q \\
L^\infty(E_0(A), \mathcal{B}(E_0(A)), \mu) & \xrightarrow[K_\mu]{} & \mathcal{C}_\mu
\end{array}$$

where q is the mapping

$$q : \mathcal{C}_\nu \ni c \mapsto ce_0 \in \mathcal{C}_\mu,$$

which is correctly defined and surjective, by virtue of the equality just obtained. Since ξ_g^0 is cyclic for $\pi_g(B)$, it is separating for $\pi_g(B)'$; hence, ξ_g^0 is separating for $\mathcal{C}_\nu$. From

150

the fact that $e_0 \xi_g^0 = \xi_g^0$, we infer that q is injective; hence, it is an isomorphism. We now easily infer that q_∞ is a surjective isomorphism. It follows that ν is a stable lifting of μ.

(b) $\Rightarrow$ (a). Indeed, assume that $[\phi] \in L^\infty(E_0(A), \mathcal{B}(E_0(A)), \mu)$ is a projection. Since μ is assumed to be orthogonal, it follows that $K_\mu([\phi])$ is a projection. Since ν is assumed to be a stable lifting of μ, from (1) we infer that $e_0 e e_0$ is a projection, for any projection $e \in \mathcal{C}_\mu$; i.e., $e_0 e e_0 e e_0 = e_0 e e_0$. From

$$(e_0 e e_0 - e e_0)^*(e_0 e e_0 - e e_0) = (e_0 e e_0 - e_0 e)(e_0 e e_0 - e e_0)$$

$$= e_0 e e_0 e e_0 - e_0 e e_0 e e_0 - e_0 e e_0 e e_0 + e_0 e e_0 = 0,$$

we infer that $e_0 e e_0 = e e_0$; hence, we have that

$$e e_0 = e_0 e,$$

for any projection $e \in \mathcal{C}_\nu$. Since linear combinations of projections are uniformly dense in $\mathcal{C}_\nu$, we infer that

$$c e_0 = e_0 c \qquad c \in \mathcal{C}_\nu;$$

hence, $e_0 \in \mathcal{C}_\nu'$, and the Theorem is proved.

IV. We shall say that a positive linear functional $f \in A_+^*$ is *simple* if $\pi_f(A)'$ is commutative. We shall denote by $S_0(A)$ the set of all simple quasi-states of A, whereas $S(A)$ will stand for the set of all simple states of A.

Theorem 20. *For any C^*-algebra A we have*
(a) $P(A) \subset S(A)$;
(b) $S(A) \cap F(A) = P(A)$;
(c) $S_0(A)$ *and* $S(A)$ *are extremal subsets of* $E_0(A)$;
(d) $S_0(A) = E_0(A)$ *if, and only if, A is commutative.*

Proof: (a) For any $p \in P(A)$ we have $\pi_p(A)' = \mathbb{C}1_{H_p}$.

(b) If $f \in S(A) \cap F(A)$, then $\pi_f(A)'$ is a commutative factor; hence, $\pi_f(A)' = \mathbb{C}1_{H_p}$. It follows that $f \in P(A)$.

(c) Let $f_0 \in S_0(A)$ and assume that

(1) $$f_0 = t f_1 + (1 - t) f_2,$$

where $0 < t < 1$ and $f_1, f_2 \in E_0(A)$.

151

From (1) we infer that $f_1 \leq (1/t)f_0$ and, therefore, there exists a $T \in \pi_{f_0}(A)'$, $0 \leq T \leq (1/t)1$, such that

$$f_1(a) = \left(\pi_{f_0}(a)T\xi^0_{f_0} | \xi^0_{f_0}\right) = \left(\pi_{f_0}(a)T^{1/2}\xi^0_{f_0} | T^{\frac{1}{2}}\xi^0_{f_0}\right),$$

for any $a \in A$. Let us denote $\xi'_f = T^{\frac{1}{2}}\xi^0_{f_0}$ and denote by e_1 the projection onto $\overline{\pi_{f_0}(A)\xi'_{f_0}} \subset H_{f_0}$. Then $e_1 \in \pi_{f_0}(A)'$ and π_{f_1} can be identified with the subrepresentation $A \ni a \mapsto \pi_{f_0}(a)e_1 \in \mathcal{L}(e_1 H_{f_0})$, whereas $\xi^0_{f_1}$ can be identified with $\xi'_{f_0} \in e_1 H_{f_0}$.

We then have

$$\pi_{f_1}(A)' = e_1\pi_{f_0}(A)'e_1 = \pi_{f_0}(A)'e_1,$$

whence we infer that $\pi_{f_1}(A)'$ is commutative; hence, $f_1 \in S_0(A)$. Similarly, $f_2 \in S_0(A)$. It follows that $S_0(A)$ is an extremal subset of $E_0(A)$. Since $E(A)$ is a face of $E_0(A)$, it immediately follows that $S(A) = S_0(A) \cap E(A)$ is an extremal subset of $E_0(A)$ (and of $E(A)$).

(d) This follows immediately from ([25], Lemma 15). The Theorem is proved.

§6. Universal Maximal Orthogonal Liftings of the Central Measures

In this section we shall prove that the central measures on $E_0(A)$ induce regular Borel measures on $F(A)$, with respect to the central topology.

I. According to Sakai's Theorem (see [24], Theorem 3.1.8; and also Theorems 17 and 18 above, for the case of an arbitrary C^*-algebra), any central Radon probability measure on $E_0(A)$, whose barycenter is a state f_0, is the stable projection of a maximal orthogonal Radon probability measure on the state space $E(B_{f_0})$ of a suitably chosen C^*-algebra B_{f_0}, with a unit element.

In order to ensure a better control of the properties of the central measures, it seems that a universal construction is better adapted to this aim.

Namely, we shall consider an arbitrary C^*-algebra A and its *universal representation* $\pi_u : A \to \mathcal{L}(H_u)$, where $H_u = \oplus_{f \in E(A)} H_u$, and $\pi_u = \oplus_{f \in E(A)} \pi_f$. Then $\pi_u(A)''$ can be canonically identified with the second dual A^{**} of A, as a Banach space, endowed with the Arens multiplication (see [12], §12; [24], §1.17, for details).

We shall denote by B the C^*-algebra $C^*(\pi_u(A), \pi_u(A)')$, generated by $\pi_u(A)$ and $\pi_u(A)'$ in $\mathcal{L}(H_u)$. Then

$$B' = \pi_u(A)' \cap \pi_u(A)''$$

is the center of $\pi_u(A)'$, of $\pi_u(A)''$ and of B itself.

Let $\pi : A \to B$ be the corestriction of π_u to B and denote by $s : E(B) \to E_0(A)$ the affine continuous mapping

$$s : E(B) \ni g \mapsto g \circ \pi \in E_0(A).$$

Let us denote by $V(B)$ the set of all vector states of the C^*-algebra $B \subset \mathcal{L}(H_u)$.

Lemma 8. (a) $V(B) \subset S(B)$; (b). $V(B)$ is an extremal subset of $E(B)$.

Proof: (a) Let $v \in V(B)$. Then there exists a $\xi \in H_u, \|\xi\| = 1$, such that $v = \omega_\xi|B$. Let us now remark that the representation $\pi_v : B \to \mathcal{L}(H_v)$ can be identified with the subrepresentation

$$B \ni b \mapsto be_\xi \in \mathcal{L}(e_\xi H_u)$$

where $e_\xi \in \mathcal{L}(H_u)$ is the projection onto $\overline{B\xi} \subset H_u$. Of course, we have that $e_\xi \in B'$; i.e., it is a central projection. It follows that $\pi_v(B)'$ can be identified with $B' e_\xi \subset \mathcal{L}(e\, H_u)$, which is an abelian von Neumann algebra; hence, $v \in S(B)$.

(b) Assume now that $v \in V(B)$ decomposes as

$$v = tv' + (1 - t)v'', \quad 0 < t < 1, \ v', \ v'' \in E(B).$$

Then $v' \leq t^{-1}v$; hence, it exists an $x \in B' e_\xi$, such that $0 \leq x \leq t^{-1}1$, and

$$v'(b) = (bx\xi|\xi) = \left(bx^{\frac{1}{2}}\xi|x^{\frac{1}{2}}\xi\right), \ b \in B.$$

We infer that $v' = \omega_{x^{\frac{1}{2}}\xi}|B$; hence, $v' \in V(B)$. Similarly, $v'' \in V(B)$, and the Lemma is proved.

Let us denote $F_0(A) = \{\lambda f; \lambda \in [0,1], f \in F(A)\}$.

Lemma 9. (a) For any $p \in F(B)$ we have that $s(p) \in F_0(A)$; (b) For any $f \in F(A)$ we have that $\omega_{\xi_f^0}|B \in P(B) \cap V(B)$ and $s(\omega_{\xi_f^0}|B) = f$.

Proof: (a) We obviously have that $\pi_p(\pi_u(A)') \subset \pi_p(\pi_u(A))'$, and therefore $\pi_p(\pi_u(A)')' \supset \pi_p(\pi_u(A))''$. From the equality

$$\pi_p(B) = C^* \left(\pi_p(\pi_u(A)), \pi_p(\pi_u(A)')\right)$$

and from the fact that $p \in F(B)$, we infer that

$$\begin{aligned}
\mathbb{C}1_{H_p} &= \pi_p(B)' \cap \pi_p(B)'' = \pi_p(\pi_u(A))' \cap \pi_p\left(\pi_u(A)'\right)' \cap \pi_p(B)'' \\
&\supset \pi_p(\pi_u(A))' \cap \pi_p(\pi_u(A))'' \cap \pi_p(\pi_u(A))'' \\
&= \pi_p(\pi_u(A))' \cap (\pi_p(\pi_u(A)))'' ;
\end{aligned}$$

hence, $\pi_p(\pi_u(A))''$ is a factor in $\mathcal{L}(H_p)$. Let us now remark that $\pi_{s(p)}$ can be identified with the subrepresentation

$$A \ni a \mapsto \pi_p(\pi_u(a))e_p \in \mathcal{L}(e_p H_p),$$

where $e_p \in \mathcal{L}(H_p)$ is the projection onto $\overline{\pi_p(\pi_u(A))\xi_p^0}$; hence, $\overset{.}{e}_p \in \pi_p(\pi_u(A))'$. It follows that $\pi_{s(p)}(A)''$ is a factor and, therefore, we have that $s(p) \in F_0(A)$.

Remark. We have $\|s(p)\| = \|e_p \xi_p^0\|^2$.

(b) Let us denote $p = \omega_{\xi_f^0}|B$. Then π_p can be identified with the representation

$$B \ni b \mapsto be_f \in \mathcal{L}(e_f H_u),$$

where $e_f \in B'$ is the projection onto $\overline{B\xi_f^0} \subset H_u$. Since $f \in F(A), e_f$ is a minimal projection of B'; hence, $B'' e_f$ is a factor. From $(B'' e_f)' = B' e_f = \mathbb{C}e_f$, we infer that $B'' e_f = \mathcal{L}(e_f H_u)$; hence, $p \in P(B)$. It is obvious that we have $p \in V(B)$, and also

$$s(p)(a) = \left(\pi_u(a)\xi_f^0|\xi_f^0\right) = f(a), \qquad a \in A.$$

The lemma is proved.

II. For any $g \in E(B)$ we have

$$\pi_g(B) = C^* \left(\pi_g\left(\pi_u(A)\right), \pi_g\left(\pi_u(A)'\right) \right)$$

and, therefore,

$$\pi_g(B)' = \pi_g\left(\pi_u(A)\right)' \cap \pi_g(\pi_u(A)')'.$$

From the fact that $\pi_g(\pi_u(A)') \subset \pi_g(\pi_u(A))'$, we infer that

$$\pi_g(B)' \supset \pi_g(\pi_u(A))' \cap \pi_g(\pi_u(A))''.$$

It is obvious that the von Neumann algebra

$$\mathcal{D}_g = \pi_g(\pi_u(A))' \cap \pi_g(\pi_u(A))''$$

is contained in the center of $\pi_g(B)'$; the corresponding orthogonal measure ν_g on $E(B)$ is, therefore, subcentral. If we denote by $e_g \in \pi_g(\pi_u(A))'$ the projection onto $\overline{\pi_g(\pi_u(A))\xi_g^0} \subset$

H_g, then the representation $\pi_{s(g)} : A \to \mathcal{L}(H_{s(g)})$ can be identified with the subrepresentation $A \ni a \mapsto \pi_g(\pi_u(a))e_g$ of A on $\mathcal{L}(e_g H_g)$. Since we have that $e_g \in \mathcal{D}_g'$, we can apply Theorem 19. From the equality

$$\mathcal{D}_g e_g = \left(e_g \pi_g \left(\pi_u(A)\right)' e_g\right) \cap \left(\pi_g \left(\pi_u(A)\right)'' e_g\right),$$

we infer that $s_*(\nu_g) = \mu_{s(g)}$ *is the central measure on* $E_0(A)$*, corresponding to* $s(g)$*, if* $s(g) \in E(A)$. Of course, ν_g is a stable lifting of $\mu_{s(g)}$.

Let us now consider the function $n : E(B) \to [0,1]$, given by $n(g) = \|s(g)\|, g \in E(B)$; it is obviously affine and lower semi-continuous. It follows that the set

$$n^{-1}(\{1\}) = E_1(B) = \{g \in E(B); s(g) \in E(A)\}$$

is a *measure extremal face and a G_δ-subset* of $E(B)$; hence, for any $\nu \in \mathcal{M}_+^1(E(B))$, such that $b(\nu) \in E_1(B)$, we have that $\nu(E_1(B)) = 1$.

We shall consider now the set $\Omega_1(E(B)) \subset \Omega(E(B))$ of all orthogonal Radon probability measures ν_0 on $E(B)$, such that
(a) $g_0 = b(\nu_0) \in E_1(B)$, and
(b) $\mathcal{C}_{\nu_0} \subset \mathcal{D}_{b(\nu_0)}$.

From (a) we infer that $b(s_*(\nu_0)) \in E(A)$, whereas from (b) we infer that $e_{b(\nu_0)} \in \mathcal{C}_{\nu_0}'$, for any $\nu_0 \in \Omega_1(E(B))$. For any $g_0 \in E_1(B)$ let $\Omega_1(E(B); g_0) \subset \Omega_1^0(E(B))$ be the set of all $\nu_0 \in \Omega_1(E(B))$, such that $b(\nu_0) = g_0$.

Lemma 10. *For any* $g \in E_1(B)$ *the restriction of* s_* *to* $\Omega_1(E(B); g)$ *is a bijection between* $\Omega_1(E(B); g)$ *and the set* $\mathcal{Z}_+^1(E_0(A); s(g))$.

Proof: By Theorem 19, since $e_g \in \mathcal{C}_\nu'$, for any $\nu \in \Omega_1(E(B); g)$, the measure $s_*(\nu)$ is orthogonal and $\nu \prec \nu_g$ implies that $s_*(\nu) \prec s_*(\nu_g) = \mu_{s(g)}$. Let us now remark that the mapping

$$s_g : \mathcal{D}_g \ni c \mapsto ce_g \in \mathcal{D}_g e_g$$

is a *-isomorphism of abelian von Neumann algebras; hence, it induces a bijection between the set of all von Neumann subalgebras of $\mathcal{D}_g$ and the set of all von Neumann subalgebras of $\mathcal{D}_g e_g = \mathcal{C}_{\mu_{s(g)}}$. Since the diagram

$$\begin{array}{ccccccc} \Omega_1(E(B); g) & \ni & \nu & \mapsto & \mathcal{C}_\nu & \subset & \mathcal{D}_g \\ s_* \downarrow & & \downarrow & & \downarrow & & \downarrow s_g \\ \mathcal{Z}_+^1(E_0(A); s(g)) \ni s_*(\nu) & \mapsto & \mathcal{C}_{s_*(\nu)} & \subset & \mathcal{C}_{\mu_{s(g)}} \end{array}$$

is commutative, the Lemma now immediately follows from ([28], Theorem 3.4).

We shall now consider the Z_1-extremal subsets $F \subset E(B)$, defined as follows: F is said to be Z_1-*extremal* if the following conditions are satisfied:

(a) F is a compact subset of $E(B)$, and

(b) $g \in F \cap E_1(B) \Rightarrow \nu(F) = 1$, for any $\nu \in \Omega_1(E(B); g)$.

It is obvious that the set $Z_1(E(B))$ of all (compact) Z_1-extremal subsets of $E(B)$ is the set of all closed subsets of a topology on $E(B)$, which we shall call the Z_1-*topology*. We shall also consider the topology $\hat{Z}_1$ induced on $s^{-1}(F(A))$ by Z_1 and we shall denote

$$\hat{Z}_1(s^{-1}(F(A))) = \{F \cap s^{-1}(F(A)); F \in Z_1(E(B))\}.$$

Let $\hat{s} : s^{-1}(F(A)) \to F(A)$ be the restriction of s to $s^{-1}(F(A))$, followed by the corresponding corestriction.

Lemma 11. *If $F_0 \subset E_0(A)$ is a compact Z-extremal subset, then $s^{-1}(F_0)$ is a compact Z_1-extremal subset of $E(B)$.*

Proof: Let $g \in s^{-1}(F_0) \cap E_1(B)$; then $s(g) \in F_0 \cap E(A)$ and, therefore, we have that $\mu(F_0) = 1$, for any subcentral measure $\mu \in M_+^1(E_0(A))$, such that $b(\mu) = s(g)$. We infer that $\nu(s^{-1}(F_0)) = s_*(\nu)(F_0) = 1$, for any $\nu \in \Omega_1(E(B), g)$, because $s_*(\nu)$ is a subcentral measure and $b(s_*(\nu)) = s(g)$. The lemma is proved.

Lemma 12. *The mapping $\hat{s}$ is a continuous surjection, if $s^{-1}(F(A))$ is endowed with the $\hat{Z}_1$-topology, whereas $F(A)$ is endowed with the central topology.*

Proof: Any centrally closed subset $\hat{F}_0 \subset F(A)$ is of the form $\hat{F}_0 = F_0 \cap F(A)$, where $F_0 \subset E_0(A)$ is a compact Z-extremal subset. Since

$$\hat{s}^{-1}(\hat{F}_0) = s^{-1}(F_0) \cap s^{-1}(F(A)),$$

it will be sufficient to prove that $s^{-1}(F_0)$ is Z_1-extremal. It is clearly compact, whereas from

$$g \in s^{-1}(F_0) \cap E_1(B)$$

we infer that $s_*(\nu) = \mu$ is a subcentral measure corresponding to $s(g)$ for any $\nu \in \Omega_1(E(B); g)$. Since $s(g) \in F_0 \cap E(A)$, we infer that $\mu(F_0) = 1$ and, therefore,

$$1 = \mu(F_0) = s_*(\nu)(F_0) = \nu(s^{-1}(F_0)),$$

for any $\nu \in \Omega_1(E(B); g)$, and the Lemma is proved.

156

Lemma 13. *Let $f \subset E(B)$ be any compact Z_1-extremal subset of $E(B)$. Then $(ex \; \overline{co}(F)) \cap E_1(B) \subset F \cap s^{-1}(F(A))$.*

Proof: By Milman's Converse Theorem, we have that $ex \; \overline{co}(F) \subset F$. Let now $g \in (ex \; \overline{co}(F)) \cap E_1(B)$. Then we have $g \in F \cap E_1(B)$ and, therefore $\nu_g(F) = 1$. We infer that $\nu_g(\overline{co}(F)) = 1$. Since $b(\nu_g) = g \in ex \; \overline{co}(F)$, by Bauer's Therem (see [23], Proposition 1.4) we infer that $\nu_g = \epsilon_g$, the Dirac measure at g. We infer that $\mathcal{D}_g = \mathbb{C}1_{H_g}$ and, therefore, $\pi_g(\pi_u(A))''$ is a factor; hence, $\pi_g(\pi_u(A))'' e_g$ is a factor, and this implies that $\pi_{s(g)}(A)''$ is a factor. From $g \in E_1(B)$ we infer that $s(g) \in F(A)$. The Lemma is proved.

Lemma 14. *Let $\nu \in \mathcal{M}_+^1(E(B))$ be any maximal measure, such that $\nu(E_1(B)) = 1$, and let $F \subset E(B)$ be any compact Z_1-extremal subset. Then*

$$F \cap s^{-1}(F(A)) = \emptyset \Rightarrow \nu(F) = 0.$$

Proof: By way of contradiction, let us assume that $\nu(F) > 0$. Define $\nu_F = \nu(F)^{-1}\chi_F \nu$. Then ν_F is a maximal Radon probability measure on $E(B)$, such that $\nu_F(F) = 1$ and

$$\nu_F(E_1(B)) = \nu(F)^{-1}\nu(F \cap E_1(B)) = 1.$$

We infer that $\nu_F(\overline{co}(F)) = 1$; hence, $\nu_F|\overline{co}(F)$ is a maximal Radon probability measure on $K = \overline{co}(F)$. If we denote $\psi = n|K$, then ψ is a lower semi-continuous affine function on K, such that $\psi(K) \subset [0,1]$ and $\nu_f(\psi) = 1$.

For any $D \in \mathcal{B}_0(E(B))$ we have $D \cap K \in \mathcal{B}_0(K)$ and also

$$(D \cap K) \cap (ex \; \psi^{-1}(\{1\})) = \emptyset,$$

by Lemma 13. From ([28], Proposition 1.8) we infer that $\nu_F(D) = 0$; i.e., $\nu_F = 0$, a contradiction. The Lemma is proved.

Lemma 15. *Let $F_0 \subset E_0(A)$ be any compact Z-extremal subset of $E_0(A)$, and let $\mu \in \mathcal{M}_+^1(E_0(A))$ be any central measure, such that $b(\mu) \in E(A)$ and $\mu(F_0) = 1$. Then we have $\mu(D) = 0$, for any $D \in \mathcal{B}_0(E_0(A))$, such that $D \cap F_0 \cap F(A) = \emptyset$.*

Proof: Let $f = b(\mu)$ and $g = \omega_{\xi_f^0}|B$. Then $\pi_g(B)' = \mathcal{D}_g$, and ν_g is a maximal orthogonal measure on $E(B)$, such that $s_*(\nu_g) = \mu$. If $D \in \mathcal{B}_0(E_0(A))$ and $D \cap F_0 \cap F(A) = \emptyset$, then $s^{-1}(D) \in \mathcal{B}_0(E(B))$, and

$$s^{-1}(D) \cap s^{-1}(F_0) \cap s^{-1}(F(A)) = \emptyset.$$

Since we have

$$\nu_g(s^{-1}(F_0)) = s_*(\nu_g)(F_0) = \mu(F_0) = 1,$$

we infer that $\nu_g(\overline{co}(s^{-1}(F_0))) = 1$.

By Lemmas 11 and 13 we have

$$\left(ex\ \overline{co}\left(s^{-1}(F_0)\right)\right) \cap E_1(B) \subset s^{-1}(F_0) \cap s^{-1}(F(A))$$

and, therefore,

$$\left(ex\ \overline{co}\left(s^{-1}(F_0)\right)\right) \cap E_1(B) \cap s^{-1}(D) = \emptyset.$$

If we denote $K = \overline{co}(s^{-1}(F_0)), \psi = n|K, \nu_0 = \nu_g|\overline{co}(s^{-1}(F_0))$, we can apply ([28], Proposition 1.8) in order to infer that $\nu_0(s^{-1}(D)\cap K) = 0$, since $s^{-1}(D)\cap K \in \mathcal{B}_0(K)$, by taking into account also the fact that $\nu_g(E_1(B)) = 1$. It follows that $\mu(D) = \nu_g(s^{-1}(D)) = 0$, and the Lemma is proved.

As an immediate consequence of Lemma 15 we obtain the following Theorem, which extends Sakai's Theorem (see [24], Theorem 3.1.8) to the case of an arbitrary C^*-algebra A, possibly not containing the unit element.

Theorem 21. *Let $\mu \in \mathcal{M}^1_+(E_0(A))$ be any central measure, such that $b(\mu) \in E(A)$. Then $\mu(D) = 0$, for any $D \in \mathcal{B}_0(E_0(A))$, such that $D \cap F(A) = \emptyset$.*

Proof: In the preceding Lemma take $F_0 = E_0(A)$.

III. We shall now consider the σ-algebra $\hat{\mathcal{B}}_0(F(A))$ defined by

$$\hat{\mathcal{B}}_0(F(A)) = \{D \cap F(A); D \in \mathcal{B}_0(E_0(A))\}.$$

Let now $\mu \in \mathcal{M}^1_+(E_0(A))$ be any central measure, such that $b(\mu) \in E(A)$. By Theorem 21 we can define correctly a probability measure

$$\hat{\mu}_0 : \hat{\mathcal{B}}_0(F(A)) \to [0,1],$$

by the formula

$$\hat{\mu}_0(D \cap F(A)) = \mu(D), \qquad D \in \mathcal{B}_0(E_0(A)).$$

From $\hat{\mu}_0$ we can derive the outer measure $\hat{\mu}_0^*$, as usually.

158

Theorem 22. *For any compact Z-extremal subset $F_0 \subset E_0(A)$ we have*

$$\hat{\mu}_0^*(F_0 \cap F(A)) = \mu(F_0).$$

Proof: Assume first that $\mu(F_0) = 0$. Then there exists a set $D_0 \in \mathcal{B}_0(E_0(A))$ such that $D_0 \supset F_0$ and $\mu(D_0) = 0$. We obviously have that $D_0 \cap F(A) \in \hat{\mathcal{B}}_0(F(A))$ and $D_0 \cap F(A) \supset F_0 \cap F(A)$.

From $\hat{\mu}_0(D_0 \cap F(A)) = \mu(D_0) = 0$ we infer that $\hat{\mu}_0^*(F_0 \cap F(A)) = 0$, and the equality is proved in this case.

If $\mu(F_0) > 0$, let us consider the central measure $\mu_{F_0} = \mu(F_0)^{-1} \chi_{F_0} \mu$. We have

$$\|b(\mu_{F_0})\| = \int_{E_0(A)} \|f\| \, d\mu_{F_0} = \mu(F_0)^{-1} \int_{F_0} \|f\| \, d\mu$$

$$= \mu(F_0)^{-1} \int_{F_0 \cap E(A)} \|f\| \, d\mu = \mu(F_0)^{-1} \mu(F_0 \cap E(A)) = 1,$$

since $\| \cdot \|$ is affine and lower semi-continuous, and $\mu(E(A)) = 1$. It follows that $b(\mu_{F_0}) \in E(A)$.

Since $\mu_{F_0}(F_0) = 1$, by Lemma 15 we have that $\mu_{F_0}(D) = 0$ for any $D \in \mathcal{B}_0(E_0(A))$, such that $D \cap F_0 \cap F(A) = \emptyset$.

Let then $D_0 \in \mathcal{B}_0(E_0(A))$ be such that

$$D_0 \cap F(A) \supset F_0 \cap F(A).$$

With $D_1 = \mathcal{C}D_0$ we have $D_1 \in \mathcal{B}_0(E_0(A))$ and $D_1 \cap F_0 \cap F(A) = \emptyset$. From Lemma 15 we infer that $\mu_{F_0}(D_1) = 0$, and this implies that $\mu_{F_0}(D_0) = 1$. It follows that

$$\mu(F_0) = \mu(F_0 \cap D_0) \le \mu(D_0) = \hat{\mu}_0(D_0 \cap F(A)),$$

and this implies that

$$\mu(F_0) \le \hat{\mu}_0^*(F_0 \cap F(A)).$$

On the other hand, there exists a $D \in \mathcal{B}_0(E_0(A))$, such that $F_0 \subset D$ and $\mu(F_0) = \mu(D)$. From $F_0 \cap F(A) \subset D \cap F(A)$, and from

$$\hat{\mu}_0^*(F_0 \cap F(A)) \le \hat{\mu}_0(D \cap F(A)) = \mu(D) = \mu(F_0),$$

we infer that $\mu(F_0) = \hat{\mu}_0^*(F_0 \cap F(A))$ and the Theorem is proved.

159

Theorem 23. *Any Baire measurable subset $M \subset F(A)$ with respect to the central topology is $\hat{\mu}_0$-measurable for any central measure $\mu \in \mathcal{M}^1_+(E_0(A))$ such that $b(\mu) \in E(A)$.*

Proof: We recall that the Baire measurable subsets of a topological space are defined to be those belonging to the smallest σ-algebra of subsets of the space, which contains all the *closed G_δ-subsets*.

It will be sufficient to prove that any centrally closed, centrally G_δ-subset $F_0 \cap F(A)$ of $F(A)$, where F_0 is a compact Z-extremal subset of $E_0(A)$, is $\hat{\mu}_0$-measurable. Indeed, if $(F_n)_{n \geq 1}$ is an increasing sequence of compact Z-extremal subsets of $E_0(A)$ such that

$$\bigcup_{n=1}^{\infty} (F_n \cap F(A)) = F(A) \backslash F_0,$$

then we have

$$\hat{\mu}_0^* (F(A) \backslash F_0) = \sup \hat{\mu}_0^* (F_n \cap F(A)).$$

From the fact that $F_n \cap F_0 \cap F(A) = \emptyset$, for $n \geq 1$, and from Theorem 10 we infer that $F_n \cap F_0 \cap E(A) = \emptyset$ and, therefore, $\mu(F_n \cap F_0) = 0$. It follows that

$$\hat{\mu}_0^* (F(A) \backslash F_0) \leq 1 - \hat{\mu}_0^* (F_0 \cap F(A)),$$

and this shows that $F_0 \cap F(A)$ is $\hat{\mu}_0$-measurabe. The Theorem is proved.

Remark. With an obvious notation, we can write the preceding result as follows:

$$\mathcal{B}_0(F(A); \hat{Z}(F(A))) \subset \hat{\mathcal{B}}_0(F(A))(\hat{\mu}_0).$$

We shall prove below that $\hat{\mu}_0$ can be extended as a regular probability measure on the σ-algebra $\mathcal{B}(F(A); \hat{Z}(F(A)))$ of all Borel measureable subsets of $F(A)$, with respect to the central topology.

We shall denote by $\hat{\mathcal{B}}_2(F(A))$ the σ-algebra of subsets of $F(A)$, generated by $\hat{\mathcal{B}}_0(F(A))$ and by $\mathcal{B}(F(A); \hat{Z}(F(A)))$. By $\mathcal{B}_2(E_0(A))$ we shall denote the σ-algebra of subsets of $E_0(A)$, generated by $Z(E_0(A))$ and $\mathcal{B}_0(E_0(A))$. It is obvious that

$$\hat{\mathcal{B}}_2(F(A)) = \{M \cap F(A); M \in \mathcal{B}_2(E_0(A))\}.$$

Also, we shall denote by $\mathcal{B}_1(E_0(A))$ the σ-algebra of subsets of $E_0(A)$ generated by the set $\mathcal{F}(E_0(A))$ of the compact extremal subsets of $E_0(A)$ and by $\mathcal{B}_0(E_0(A))$. From the inclusion $\mathcal{F}(E_0(A)) \subset Z(E_0(A))$ we immediately infer that $\mathcal{B}_1(E_0(A)) \subset \mathcal{B}_2(E_0(A))$.

Lemma 16. *Let $\nu \in \mathcal{M}^1_+(E(B))$ be a maximal measure such that $\nu(E_1(B)) = 1$. Then $\nu(D) = 0$ for any $D \in \mathcal{B}_0(E(B))$ such that $D \cap s^{-1}(F(A)) = \emptyset$.*

Proof: From ([31], Theorem 2) we infer that

$$\widetilde{\nu}(P(B) \cap E_1(B)) = 1.$$

(Here $\widetilde{\nu}$ is the C-Borel probability measure induced on $P(B)$ by ν; see [31].)

The inclusion $P(B) \cap E_1(B) \subset s^{-1}(F(A))$ now implies that

$$\nu(D) = \widetilde{\nu}(D \cap P(B)) = 0,$$

for any $D \in \mathcal{B}_0(E(B))$ such that $D \cap s^{-1}(F(A)) = \emptyset$. The Lemma is proved.

We can now consider the σ-algebra $\hat{\mathcal{B}}_0(s^{-1}(F(A)))$ defined by

$$\hat{\mathcal{B}}_0\left(s^{-1}(F(A))\right) = \left\{D \cap s^{-1}(F(A)); D \in \mathcal{B}_0(E(B))\right\}.$$

The preceding lemma shows that by the formula

$$\hat{\nu}_0\left(D \cap s^{-1}(F(A))\right) = \nu(D), \qquad D \in \mathcal{B}_0(E(B)),$$

one defines correctly a probability measure

$$\hat{\nu}_0 : \hat{\mathcal{B}}_0\left(s^{-1}(F(A))\right) \to [0,1],$$

from which one can derive the corresponding outer measure ν_0^*.

Lemma 17. *Let $F \subset E(B)$ be any compact Z_1-extremal subset of $E(B)$, and let $\nu \in \mathcal{M}^1_+(E(B))$ be any maximal measure, such that $\nu(E_1(B)) = 1$ and $\nu(F) = 1$. Then we have $\nu(D) = 0$ for any $D \in \mathcal{B}_0(E(B))$ such that $D \cap F \cap s^{-1}(F(A)) = \emptyset$.*

Proof: Let $K = \overline{co}(F)$; then $\nu(K) = 1$ and $\nu|K$ is a maximal Radon probability measure on K. If $D \in \mathcal{B}_0(E(B))$ then $D \cap K \in \mathcal{B}_0(K)$. If we denote $\psi = n|K$, then ψ is a lower semi-continuous affine function $\psi : K \to [0,1]$ such that $\nu(\psi) = 1$. Since $K \cap E_1(B) = \psi^{-1}(\{1\})$, from the fact that $ex\ \psi^{-1}(\{1\}) = E_1(B) \cap (ex\ K)$ from Lemma 13 and from ([28], Proposition 1.8) the present Lemma now immediately follows.

Theorem 24. *For any compact Z_1-extremal subset $F \subset E(B)$ we have*

$$\hat{\nu}_0^*(F \cap s^{-1}(F(A))) = \nu(F).$$

161

(Here $\nu \in \mathcal{M}_+^1(E(B))$ is any maximal measure, such that $\nu(E_1(B)) = 1$.)

Proof: If $\nu(F) = 0$, then there exists a $D \in \mathcal{B}_0(E(B))$ such that $F \subset D$ and $\nu(D) = 0$. We then have

$$F \cap s^{-1}(F(A)) \subset D \cap s^{-1}(F(A))$$

and therefore,

$$\hat{\nu}_0^* \left(F \cap s^{-1}(F(A)) \right) \leq \hat{\nu}_0 \left(D \cap s^{-1}(F(A)) \right) = \nu(D) = 0.$$

Let us now assume that $\nu(F) > 0$. We shall then define the maximal measure $\nu_F = \nu(F)^{-1} \chi_F \nu$ for which we also have $\nu_F(E_1(B)) = 1$. By Lemma 17 we have that $\nu_F(D) = 0$ for any $D \in \mathcal{B}_0(E(B))$ such that $D \cap F \cap s^{-1}(F(A)) = \emptyset$ because $\nu_F(F) = 1$.

Let then $D_0 \in \mathcal{B}_0(E(B))$ be such that

$$D_0 \cap s^{-1}(F(A)) \supset F \cap s^{-1}(F(A)).$$

With $D_1 = \mathcal{C}D_0$ we have that $D_1 \in \mathcal{B}_0(E(B))$ and $D_1 \cap F \cap s^{-1}(F(A)) = \emptyset$. From Lemma 17 we now infer that $\nu_F(D_1) = 0$ and this implies that $\nu_F(D_0) = 1$. It follows that

$$\nu(F) = \nu(F \cap D_0) \leq \nu(D_0) = \hat{\nu}_0(D_0 \cap s^{-1}(F(A))),$$

and this implies that

$$\nu(F) \leq \hat{\nu}_0^*(F \cap s^{-1}(F(A))).$$

On the other hand, there exists a $D \in \mathcal{B}_0(E(B))$ such that $F \subset D$ and $\nu(F) = \nu(D)$. From $F \cap s^{-1}(F(A)) \subset D \cap s^{-1}(F(A))$ and from

$$\hat{\nu}_0^* \left(F \cap s^{-1}(F(A)) \right) \leq \hat{\nu}_0 \left(D \cap s^{-1}(F(A)) \right) = \nu(D) = \nu(F),$$

we infer that $\nu(F) = \hat{\nu}_0^*(F \cap s^{-1}(F(A)))$ and the Theorem is proved.

IV. We shall now use the preceding results in order to prove that any central measure $\mu \in \mathcal{M}_+^1(E_0(A))$ such that $b(\mu) \in E(A)$ induces a regular Borel measure on $F(A)$ with respect to the central topology on $F(A)$. To this end we shall adapt to our case the method of proof given by Batty to a similar problem (see [5], Theorem 7; [6], Theorem 3.2; [30], Theorem 13).

Let $\mathcal{B}_2(E(B))$ be the σ-algebra of subsets of $E(B)$ generated by $\mathcal{B}_0(E(B))$ and by the set of all (compact) Z_1-extremal subsets of $E(B)$. Of course, any compact extremal subset of $E(B)$ belongs to $\mathcal{B}_2(E(B))$. We shall also consider the σ-algebra $\hat{\mathcal{B}}_2(s^{-1}(F(A)))$ of subsets of $s^{-1}(F(A))$ given by

$$\hat{\mathcal{B}}_2\left(s^{-1}(F(A))\right) = \left\{M \cap s^{-1}(F(A)); M \in \mathcal{B}_2(E(B))\right\}.$$

It is clear that we have the inclusions

$$\mathcal{B}_0(E(B)) \subset \mathcal{B}_1(E(B)) \subset \mathcal{B}_2(E(B)) \subset \mathcal{B}(E(B)),$$

where by $\mathcal{B}_1(E(B))$ we have denoted the σ-algebra of subsets of $E(B)$ generated by $\mathcal{B}_0(E(B))$ and by the set $\mathcal{F}(E(B))$ of all compact extremal subsets of $E(B)$. Moreover, it is easy to see that $\hat{\mathcal{B}}_2(s^{-1}(F(A)))$ is the σ-algebra of subsets of $s^{-1}(F(A))$ generated by $\hat{\mathcal{B}}_0(s^{-1}(F(A)))$ and $\mathcal{B}(s^{-1}(F(A)); \hat{Z}_1)$.

Let $\mu \in \mathcal{M}^1_+(E_0(A))$ be any central measure such that $f = b(\mu) \in E(A)$, let $g = \omega_{\xi^0_f}|B$ and consider the corresponding maximal orthogonal measure $\nu = \nu_g$.

For any $M \in \mathcal{B}_2(E(B))$ we shall define

$$\nu'(M) = \sup\{\nu(F); F \in Z_1(E(B)), F \cap s^{-1}(F(A)) \subset M\}$$

and

$$\nu''(M) = \sup\{\nu(F); F \in Z_1(E(B)), F \subset M\}.$$

We have the following properties:

(a) $\nu''(M) \leq \nu'(M)$ for any $M \in \mathcal{B}_2(E(B))$; obvious.

(b) $\nu''(M) \leq \nu(M)$ for any $M \in \mathcal{B}_2(E(B))$; obvious.

(c) $\nu''(D) = \nu(D)$ for any $D \in \mathcal{B}_0(E(B))$. Indeed, it is clear that $\nu''(D) \leq \nu(D)$ by assertion (b). Since ν is maximal, and since $Z_1(E(B))$ includes all compact extremal subsets of $E(B)$, the equality follows from ([29], Corollary to Theorem 1).

(d) $\nu''(E(B)\backslash F) = \nu(E(B)\backslash F)$ for any $F \in Z_1(E(B))$. Indeed, since $E(B)\backslash F$ is open in $E(B)$, and since ν is maximal by ([29], Theorem 2) for any $\epsilon > 0$, there exists a compact extremal subset $F_1 \subset E(B)\backslash F$ such that

$$\nu(E(B)\backslash F) - \epsilon < \nu(F_1)$$

and therefore, we have

$$\nu(E(B)\backslash F) \leq \nu''(E(B)\backslash F).$$

163

Assertion (b) provides the reversed inequality.

(e) $\nu''(F) = \nu(F)$ for any $F \in Z_1(E(B))$; obvious.

(f) $\nu''(M) = \nu(M)$ for any $M \in \mathcal{B}_2(E(B))$. Indeed, as in ([5], proof of Proposition 5) we shall consider the set

$$\mathcal{B}'' = \left\{ M \in \mathcal{B}_2(E(B)); \nu''(M) = \nu(M), \nu''(\mathcal{C}M) = \nu(\mathcal{C}M) \right\}.$$

It is easy to prove that $\mathcal{B}''$ is a σ-algebra, such that

$$\mathcal{B}_0(E(B)) \subset \mathcal{B}'' \text{ and } Z_1(E(B)) \subset \mathcal{B}'',$$

by virtue of assertions (c), (d), and (e). It follows that $\mathcal{B}'' = \mathcal{B}_2(E(B))$.

(g) We have

$$\nu'(M_1) + \nu'(M_2) \leq \nu'(M_1 \cup M_2)$$

and

$$\nu''(M_1) + \nu''(M_2) \leq \nu''(M_1 \cup M_2),$$

for any $M_1, M_2 \in \mathcal{B}_2(E(B))$ such that $M_1 \cap M_2 = \emptyset$. Indeed, the second inequality is an immediate consequence of the definition of ν''; for the first, given $\epsilon > 0$, there exist $F_1, F_2 \in Z_1(E(B))$ such that

$$F_i \cap s^{-1}(F(A)) \subset M_i \text{ and } \nu'(M_i) - \epsilon < \nu(F_i), i = 1, 2.$$

We infer that

$$\nu'(M_1) + \nu'(M_2) - 2\epsilon < \nu(F_1) + \nu(F_2),$$

whereas from $F_1 \cap F_2 \cap s^{-1}(F(A)) = \emptyset$ and from Lemma 14 we infer that $\nu(F_1) + \nu(F_2) = \nu(F_1 \cup F_2)$. The assertion now immediately follows.

(h) $\nu'(M) = \nu(M)$ for any $M \in \mathcal{B}_2(E(B))$. Indeed, by (a) and (f) we have $\nu(M) \leq \nu'(M)$ for any $M \in \mathcal{B}_2(E(B))$. By (g) we have

$$1 = \nu(M) + \nu(\mathcal{C}M) \leq \nu'(M) + \nu'(\mathcal{C}M) \leq 1;$$

hence $\nu'(M) = \nu(M)$ for any $M \in \mathcal{B}_2(E(B))$.

(i) For any $M \in \mathcal{B}_2(E(B))$ we have that

$$M \cap s^{-1}(F(A)) = \emptyset \Rightarrow \nu(M) = 0.$$

Indeed, this is an immediate consequence of assertion (h).

i) We infer that by the formula

$$\hat{\nu}\left(M \cap s^{-1}(F(A))\right) = \nu(M), \qquad M \in \mathcal{B}_2(E(B)),$$

we define correctly a probability measure

$$\hat{\nu} : \hat{\mathcal{B}}_2\left(s^{-1}(F(A))\right) \to [0,1],$$

such that $\hat{\nu}|\hat{\mathcal{B}}_0(s^{-1}(F(A))) = \hat{\nu}_0$.

By summarizing the preceding results, we get the following

Theorem 25. *There exists a probability measure*

$$\hat{\nu} : \hat{\mathcal{B}}_2\left(s^{-1}(F(A))\right) \to [0,1],$$

such that
(a) $\hat{\nu}|\hat{\mathcal{B}}_0(s^{-1}(F(A))) = \hat{\nu}_0$; *and*
(b) $\hat{\nu}(M \cap s^{-1}(F(A))) = \nu(M), M \in \mathcal{B}_2(E(B))$;

and possessing the following regularity properties

(a) $\hat{\nu}(\hat{M}) = \sup\{\hat{\nu}(\hat{F}); \hat{F} \subset \hat{M}, \hat{F} \in \hat{Z}_1(s^{-1}(F(A)))\}, \hat{M} \in \hat{\mathcal{B}}_2(s^{-1}(F(A)))$; *and*
(b) $\hat{\nu}(\hat{F}) = \inf\{\hat{\nu}(\hat{D}); \hat{D} \supset \hat{F}, \hat{D} \in \hat{\mathcal{B}}_0(s^{-1}(F(A)))\}, \hat{F} \in \hat{Z}_1(s^{-1}(F(A)))$.

Proof: The existence of $\hat{\nu}$ was established just before the statement of the Theorem; assertion (a) follows from the equality $\nu' = \nu$, whereas assertion (b) follows from Theorem 24.

Let us now recall that by Lemma 12 the mapping

$$\hat{s} : s^{-1}(F(A)) \to F(A)$$

is continuous if $s^{-1}(F(A))$ is equipped with the $\hat{Z}_1$-topology, whereas $F(A)$ is equipped with the Z-topology (i.e., the central topology). Also, since $s : E(B) \to E_0(A)$ is continuous, we have that

$$D_0 \in \mathcal{B}_0(E_0(A)) \Rightarrow s^{-1}(D_0) \in \mathcal{B}_0(E(B))$$

and, therefore, we have that

$$\hat{s}_*\left(\hat{\mathcal{B}}_0\left(s^{-1}(F(A))\right)\right) \supset \hat{\mathcal{B}}_0(F(A))$$

and

$$\hat{s}_* \left(\mathcal{B} \left(s^{-1}(F(A)) \right) ; \hat{Z}_1 \right) \supset \mathcal{B}(F(A); Z),$$

where, in the left hand members of these relations the full direct images of the corresponding σ-algebras are denoted. We infer that

$$\hat{s}_* \left(\hat{\mathcal{B}}_2 \left(s^{-1}(F(A)) \right) \right) \supset \hat{\mathcal{B}}_2(F(A)),$$

and, therefore, the full direct image of the measure $\hat{\nu}$, i.e.,

$$\hat{s}_*(\nu) : \hat{s}_* \left(\hat{\mathcal{B}}_2 \left(s^{-1}(F(A)) \right) \right) \to [0, 1],$$

is defined on $\hat{\mathcal{B}}_2(F(A))$. It is easy to see that

$$\hat{s}_*(\nu) | \hat{\mathcal{B}}_0(F(A)) = \hat{\mu}_0.$$

We shall denote by $\hat{\mu}$ the restriction of $\hat{s}_*(\nu)$ to $\hat{\mathcal{B}}(F(A))$.

Lemma 18. *The mapping* $\hat{s} : s^{-1}(F(A)) \to F(A)$ *is closed.*

Proof: We must prove that

$$\hat{F} \in \hat{Z}_1 \left(s^{-1}(F(A)) \right) \Rightarrow \hat{s}(\hat{F}) \in \hat{Z}(F(A)).$$

Indeed, let $\hat{F} = F \cap s^{-1}(F(A))$ where $F \subset E(B)$ is a compact Z_1-extremal subset. Since we have

$$\hat{s}(\hat{F}) = s(F) \cap F(A),$$

it will be sufficient to prove that $s(F) \subset E_0(A)$ is Z-extremal. Indeed, for any $f_0 \in s(F) \cap E(A)$ let $\mu_0 \in \Omega(E_0(A); f_0)$ be any subcentral measure such that $b(\mu_0) = f_0$. Let $g_0 \in F$ be such that $f_0 = s(g_0)$. Then we have that $g_0 \in E_1(B)$ and, by Lemma 10, there exists a measure $\nu_0 \in \Omega_1(E(B); g_0)$ such that $s_*(\nu_0) = \mu_0$. From $b(\nu_0) = g_0 \in F$ we infer that $\nu_0(F) = 1$ and therefore we have that

$$\mu_0(s(F)) = \nu_0 \left(s^{-1}(s(F)) \right) \geq \nu_0(F) = 1,$$

and the Lemma is proved.

Remark. The proof of the Lemma shows that we have the implication

$$F \in Z_1(E(B)) \Rightarrow s(F) \in Z(E_0(A)),$$

which will be used below.

We can now prove

Theorem 26. For any central measure $\mu \in \Omega(E_0(A))$ such that $b(\mu) \in E(A)$ the corresponding measure $\hat{\mu}$ has the following properties

(a) $\hat{\mu}|\hat{\mathcal{B}}_0(F(A)) = \hat{\mu}_0$; and

(b) $\hat{\mu}(M_0 \cap F(A)) = \mu(M_0)$, for any $M_0 \in \mathcal{B}_2(E_0(A))$; as well as the following regularity properties

(c) $\hat{\mu}(\hat{M}_0) = \sup\{\hat{\mu}(\hat{F}_0); \hat{F}_0 \subset \hat{M}_0, \hat{F}_0 \in \hat{Z}(F(A))\}$ for any $\hat{M}_0 \in \hat{\mathcal{B}}_2(F(A))$; and

(d) $\hat{\mu}(\hat{F}_0) = \inf\{\hat{\mu}(\hat{D}_0); \hat{D}_0 \supset \hat{F}_0, \hat{D}_0 \in \hat{\mathcal{B}}_0(F(A))\}$, for any $\hat{F}_0 \in \hat{Z}(F(A))$.

Proof: Assertions (a) and (b) have been established just before the statement of the Theorem.

(c) Let $\hat{M}_0 \in \hat{\mathcal{B}}(E_0(A))$ and $\epsilon > 0$ be given. Then we have $\hat{s}^{-1}(\hat{M}_0) \in \hat{\mathcal{B}}_2(s^{-1}(F(A)))$. By Theorem 25 (a) there exists a $\hat{F} \in \hat{Z}_1(s^{-1}(F(A)))$, such that $\hat{F} \subset \hat{s}^{-1}(\hat{M}_0)$ and

$$\hat{\mu}(\hat{M}_0) - \epsilon = \hat{\nu}(\hat{s}^{-1}(M_0)) - \epsilon < \hat{\nu}(\hat{F}) \le \hat{\mu}(\hat{s}(\hat{F})).$$

Of course, there exists a compact Z_1-extremal subset $F \subset E(B)$ such that $\hat{F} = F \cap s^{-1}(F(A))$. Since by Lemma 18 we have that $\hat{s}(\hat{F}) \in \hat{Z}(F(A))$, the assertion now immediately follows.

(d) This is an immediate consequence of Theorem 22. The Theorem is proved.

The regularity property (c) implies that the *Borel restriction* of $\hat{\mu}$, i.e., the measure $\hat{\mu}|\mathcal{B}(F(A); \hat{Z}(F(A)))$, determines $\hat{\mu}$ on $\hat{\mathcal{B}}_2(F(A))$, as the following theorem shows.

Theorem 27. (a) $\hat{\mathcal{B}}_0(F(A)) \subset \mathcal{B}(F(A); \hat{Z}(F(A)))(\hat{\mu})$, and

(b) $\hat{\mathcal{B}}_2(F(A)) \subset \mathcal{B}(F(A); \hat{Z}(F(A)))(\hat{\mu})$.

Proof: The two statements are immediate consequences of property (c) in Theorem 26.

V. Property (c) in the statement of Theorem 26 could be called *the regularity by closed subsets*. A more careful analysis of the situation will show that we have, in fact, the stronger property of *the regularity by closed quasi-compact subsets*, which we shall present below. In fact, let us return to the proof of statement (c) from Theorem 26. Let us first remark that $s^{-1}(F(A)) \subset E_1(B)$. On the other hand, since $b(\nu) \in E_1(B)$, we have that $\nu(E_1(B)) = 1$. Since $E_1(B)$ is a G_δ-subset of $E(B)$, whereas ν is a maximal (orthogonal) measure on $E(B)$, there exists a compact extremal subset $F_1 \subset E_1(B)$ such that $\nu(F_1) > 1 - \epsilon$. Then F_1 is also Z_1-extremal and, therefore, $s(F \cap F_1)$ is compact and Z-extremal in $E_0(A)$. Since $s(F \cap F_1) \subset E(A)$, it is easy to show, with the help of

Theorem 10, that the set $s(F \cap F_1) \cap F(A)$ is $\hat{Z}(F(A))$-quasi-compact. Moreover, we have that

$$\hat{\mu}(\hat{M}_0) - 2\epsilon < \hat{\mu}(\hat{s}(\hat{F} \cap \hat{F}_1)).$$

Thus we have obtained the following

Theorem 28. *For any central measure* $\mu \in \Omega(E_0(A))$ *such that* $b(\mu) \in E(A)$, *any* $\hat{M}_0 \in \hat{B}_2(F(A))$ *and any* $\epsilon > 0$, *there exists a* $\hat{Z}$-*closed,* $\hat{Z}$-*quasi-compact set* $\hat{F}_0 \subset \hat{M}_0$ *such that* $\hat{\mu}(\hat{M}_0) - \epsilon < \hat{\mu}(\hat{F}_0)$.

Of course, this Theorem extends the well-known fact that on (Hausdorff) locally compact spaces bounded Radon measures are regular by compact subsets.

The following theorem will be used below.

Theorem 29. (a) $\quad s_*(\mathcal{B}_2(E(B))(\nu)) = \mathcal{B}_2(E_0(A))(\mu);$ and

(b) $\hat{s}_*(\hat{B}_2(s^{-1}(F(A)))(\hat{\nu})) = \hat{B}_2(F(A))(\hat{\mu})$.

Proof: (a) Since $s : E(B) \to E_0(A)$ is $(\mathcal{B}_2(E(B)); \mathcal{B}_2(E_0(A)))$-measurable, it immediately follows that

$$\mathcal{B}_2(E_0(A))(\mu) \subset s_*(\mathcal{B}_2(E(B))(\nu)).$$

Let now $M \in s_*(\mathcal{B}_2(E(B))(\nu))$. Then, by the definition of the full direct image, we have that $s^{-1}(M) \in \mathcal{B}_2(E(B))(\nu)$. By Theorem 25 and the proof preceding it (since $\nu'' = \nu$), there exist increasing sequences $(F_n)_{n \geq 0}$ and $(F'_n)_{n \geq 0}$ such that $F_n \in Z_1(E(B)), F_n \subset F_{n+1} \subset s^{-1}(M)$, for any $n \geq 0$, and $\nu(F_n) \uparrow \nu(s^{-1}(M))_n, F'_n \in Z_1(E(B)), F'_n \subset F'_{n+1} \subset Cs^{-1}(M)$ for any $n \geq 0$ and $\nu(F'_n) \uparrow \nu(s^{-1}(M))$.

By the remark following the proof of Lemma 18 we have that $s(F_n), s(F'_n) \in Z(E_0(A)), n \geq 0$ and we have that

$$s^{-1}(s(F_n)) \subset s^{-1}(s(F_{n+1})) \subset s^{-1}(M), \quad n \geq 0,$$
$$s^{-1}(s(F'_n)) \subset s^{-1}(s(F'_{n+1})) \subset Cs^{-1}(M), \quad n \geq 0,$$

and $\nu(s^{-1}(s(F_n))) \uparrow \nu(s^{-1}(M)), \nu(s^{-1}(s(F'_n))) \uparrow \nu(s^{-1}(M))$. We infer that we have

$$s(F_n) \subset s(F_{n+1}) \subset M, \quad s(F'_n) \subset s(F'_{n+1}) \subset CM, \quad n \geq 0,$$

and $\mu(s(F_n)) \uparrow \mu(M), \mu(s(F'_n)) \uparrow \mu(CM)$; it follows that $M \in \mathcal{B}_2(E_0(A))(\mu)$.

(b) Similar proof to that of (a), with the help of the fact that the mapping $\hat{s} = s|s^{-1}(F(A))$ is $(\hat{B}_2(s^{-1}(F(A)))$; $\hat{B}_2(F(A)))$-measurable, by taking into account also Lemma 18 and Theorem 26.

From the regularity property we can easily obtain the following extension of Lusin's Theorem in which we keep the preceding notation.

Theorem 30. *For any $\hat{\mu}$-measurable function $\phi : F(A) \to \mathbb{C}$, and any $\epsilon > 0$, there exists a $\hat{Z}$-closed $\hat{Z}$-quasi-compact subset $\hat{F} \subset F(A)$ such that $\hat{\mu}(\hat{F}) > 1 - \epsilon$ and $\phi|\hat{F}$ be continuous.*

Proof: Similar to that of ([33], Theorem 1.5).

Another important property of the measures $\hat{\mu}$ is exhibited by the following Theorem in which $\hat{\mu}$ is, as above, the measure induced on $\mathcal{B}(F(A); \hat{Z}(F(A)))$ by any central measure $\mu \in \mathcal{M}^1_+(E_0(A))$ such that $b(\mu) \in E(A)$.

Theorem 31. *Any measure $\hat{\mu}$ is perfect.*

Proof: Similar to that of ([33], Theorem 1.6).

§7. Central Reduction

In this section we shall develop a spatial central disintegration (reduction) theory for an arbitrary cyclic representation $\pi : A \to \mathcal{L}(H)$. We shall then show that the disintegration can be extended to the Borel enveloping C^*-algebra $\mathcal{B}(A)$ of A.

I. Let $f_0 \in E(A)$ and let $\mu = \mu_{f_0}$ be the corresponding central measure in $\Omega(E_0(A))$. Then, by Theorem 21, we have an induced probability measure $\hat{\mu}_0 : \hat{\mathcal{B}}_0(F(A)) \to [0,1]$ whereas by Theorem 27 we have a Borel extension $\hat{\mu} : \mathcal{B}(F(A); \hat{Z}(F(A)))(\hat{\mu}) \to [0,1]$ with good regularity properties such that $\hat{\mu}_0 = \hat{\mu}|\hat{\mathcal{B}}_0(F(A))$. We shall aslo denote the measure $\hat{\mu}$ by $\hat{\mu}_{f_0}$ since it is uniquely determined by $f_0 \in E(A)$.

In order to carry out the central disintegration of $\pi_{f_0} : A \to \mathcal{L}(H_{f_0})$ the measure $\hat{\mu}_0$ would be sufficient, but if we want to disintegrate over $F(A)$ the restriction of π''_{f_0} to the Borel enveloping C^*-algebra $\mathcal{B}(A)$ of A (see [36], for the definition of $\mathcal{B}(A); \pi''_{f_0} : A^{**} \to \mathcal{L}(H_{f_0})$ is the canonical extension of π_{f_0} to A^{**}), then the measure $\hat{\mu}_0$ does not suffice any more, whereas $\hat{\mu}$ will do the job, as we shall see below.

Remark. Of course, it would be senseless to try to disintegrate over $F(A)$ the representation π''_{f_0} since the algebra A^{**} is too big, as shown by the commutative case. In other cases, however, as for instance that of the elementary C^*-algebras $A = \mathcal{K}(H)$ in which $F(A) = E(A), \pi''_{f_0}$ can be disintegrated over $F(A)$, but these are the exceptions.

Lemma 19. (a) *Let $\phi : E_0(A) \to \mathbb{C}$ be any Baire measurable complex function. Then $\phi|F(A)$ is $\hat{\mathcal{B}}_0(F(A))$-measurable.* (b) *If $\phi : E_0(A) \to \mathbb{C}$ is $\mathcal{B}_2(E_0(A))$-measurable, then*

169

$\phi | F(A)$ is $\hat{\mathcal{B}}_2(F(A))$-measurable. In either case, if ϕ is μ-integrable, then $\phi | F(A)$ is $\hat{\mu}$-integrable and

$$\int_{E_0(A)} \phi \ d\mu = \int_{F(A)} \phi \ d\hat{\mu}.$$

Proof: Similar to that of ([28], Lemma 1.1) by approximating by elementary functions.

II. As in ([28], §4), we shall consider the field of Hilbert spaces $E_0(A) \ni f \mapsto H_f$ which we shall denote by $(H_f)_{f \in E_0(A)}$. We have also the associated field of triples $E_0(A) \ni f \mapsto (\pi_f, H_f, \xi_f^0)$ corresponding to the GNS-construction. If we define $\theta_f : A \to H_f$ by $\theta_f(a) = \pi_f(a)\xi_f^0, a \in A, f \in E_0(A)$, then we have a linear mapping $\theta : A \to \Pi_{f \in E_0(A)} H_f$ given by $\theta(a) = (\theta_f(a))_{f \in E_0(A)}, a \in A$. Let $\Gamma_0(A) = im \ \theta$, then $\Gamma_0(A)$ is a vector subspace of $\Pi_{f \in E_0(A)} H_f$.

As in ([28], §4), we can consider the L^2-completion $\Gamma^2(A; \mu)$ of $\Gamma_0(A)$. Of course, $\Gamma^2(\Lambda; \mu)$ is a vector subspace of $\Pi_{f \in E_0(A)} H_f$.

Since the measure μ is orthogonal, the system

(0) $$((H_f)_{f \in E_0(A)}, \Gamma^2(A; \mu), \mu)$$

is an integrable field of Hilbert spaces, in the sense of W. Wils (see [28], Theorem 4.1). Of course, the scalar product in $\Gamma_0(A)$ is given by the formula

$$((\theta_f(a))_{f \in E_0(A)} | (\theta_f(b))_{f \in E_0(A)}) = \int_{E_0(A)} f(b^*a) \ d\mu(f),$$

for any $a, b \in A$, whereas for two vector fields $(\xi_f)_{f \in E_0(A)}, (\eta_f)_{f \in E_0(A)} \in \Gamma^2(A; \mu)$, it is given by the formula

$$((\xi_f)_{f \in E_0(A)} | (\eta_f)_{f \in E_0(A)}) = \int_{E_0(A)} (\xi_f | \eta_f)_f \ d\mu(f).$$

Endowed with the corresponding semi-norm, $\Gamma^2(A; \mu)$ is a possibly non-Hausdorff, but complete pre-Hilbert space. Moreover, since μ is orthogonal, for any bounded Borel measurable function $\phi : E_0(A) \to \mathbf{C}$ we have the implication

$$(\xi_f)_{f \in E_0(A)} \in \Gamma^2(A; \mu) \Rightarrow (\phi(f)\xi_f)_{f \in E_0(A)} \in \Gamma^2(A; \mu).$$

We shall denote by $\widetilde{\Gamma}^2(A; \mu)$ the associated (Hausdorff, complete) Hilbert space, and by $\rho_\mu : \Gamma^2(A; \mu) \to \widetilde{\Gamma}^2(A; \mu)$ the corresponding canonical surjection.

We shall also consider the *restricted* field of Hilbert spaces $F(A) \ni f \mapsto H_f$ which we shall denote by $(H_f)_{f \in F(A)}$.

We can now define the linear mapping $\hat{\theta} : A \to \Pi_{f \in F(A)} H_f$ given by

$$\hat{\theta}(a) = (\theta_f(a))_{f \in F(A)}, a \in A.$$

Let us define $\hat{\Gamma}_0(A) = im \ \hat{\theta}$. Then $\hat{\Gamma}_0(A)$ is a vector subspace of $\Pi_{f \in F(A)} H_f$ and we have

$$(1) \qquad f_0(a_2^* a_1) = \int_{E_0(A)} f(a_2^* a_1) d\mu(f) = \int_{F(A)} f(a_2^* a_1) d\hat{\mu}(f),$$

for any $a_1, a_2 \in A$ by virtue of Lemma 19. We can, therefore, define a scalar product on $\hat{\Gamma}_0(A)$ by the formula

$$(2) \qquad \begin{aligned} \left(\hat{\theta}(a_1) | \hat{\theta}(a_2) \right) &= \int_{F(A)} f(a_2^* a_1) d\hat{\mu}(f) \\ &= \int_{F(A)} (\theta_f(a_1) | \theta_f(a_2)) d\hat{\mu}(f), \quad a_1, a_2 \in A. \end{aligned}$$

We can consider the *restriction mapping* $\rho : \Gamma_0(A) \to \hat{\Gamma}_0(A)$ given by $\rho(\theta(a)) = \hat{\theta}(A), a \in A$, which is a unitary linear surjection, by formulas (1) and (2).

Let $\hat{\Gamma}^2(A; \hat{\mu})$ be the L^2-completion of $\hat{\Gamma}_0(A)$ with respect to the measure $\hat{\mu}$ constructed as in ([28], §4). Then $\hat{\Gamma}^2(A; \hat{\mu})$ is a vector subspace of $\Pi_{f \in F(A)} H_f$ on which the scalar product

$$(2') \qquad \left((\xi_f)_{f \in F(A)} | (\eta_f)_{f \in F(A)} \right) = \int_{F(A)} (\xi_f | \eta_f)_f d\hat{\mu}(f),$$

for $(\xi_f)_{f \in F(A)}, (\eta_f)_{f \in F(A)} \in \hat{\Gamma}^2(A; \hat{\mu})$, is defined, along with the associated semi-norm $\| \cdot \|$, given by

$$(2'') \qquad \left\| (\xi_f)_{f \in F(A)} \right\|^2 = \int_{F(A)} \| \xi_f \|_f^2 \ d\hat{\mu}(f),$$

for any $(\xi_f)_{f \in F(A)} \in \Gamma^2(A; \hat{\mu})$.

Proposition 4.2 from [28] implies that $\hat{\Gamma}^2(A; \hat{\mu})$ is a complete, possibly non-Hausdorff, pre-Hilbert space.

The restriction mapping ρ extends to the retriction mapping $\rho_\mu : \Gamma^2(A; \mu) \to \hat{\Gamma}^2(A; \hat{\mu})$ given by $\rho_\mu((\xi_f)_{f \in E_0(A)}) = (\xi_f)_{f \in F(A)}$, for $(\xi_f)_{f \in E_0(A)} \in \Gamma^2(A; \mu)$. It is obvious that we have

$$(3) \qquad \int_{E_0(A)} (\xi_f | \eta_f)_f \ d\mu(f) = \int_{F(A)} (\xi_f | \eta_f)_f \ d\hat{\mu}(F),$$

for any $(\xi_f)_{f \in E_0(A)}, (\eta_f)_{f \in E_0(A)} \in \Gamma^2(A; \mu)$. It immediately follows that we have the implication

(4) $\qquad (\xi_f)_{f \in F(A)} \in \hat{\Gamma}^2(A; \hat{\mu}) \Rightarrow (\phi(f)\xi_f)_{f \in F(A)} \in \hat{\Gamma}^2(A; \hat{\mu}),$

for any bounded $\hat{\mu}$-measurable function $\phi : F(A) \to \mathbb{C}$ (by stability properties, any $\hat{\mu}$-measurable function on $F(A)$ coincides $\hat{\mu}$-a.e. with the restriction to $F(A)$ of a Baire measurable function on $E_0(A)$). From (4) we immediately infer that the system

$$((H_f)_{f \in F(A)}, \hat{\Gamma}^2(A; \hat{\mu}), \hat{\mu})$$

is an integrable field of Hilbert spaces, in the sense of W. Wils.

Let $q : \Gamma^2(A; \mu) \to \widetilde{\Gamma}^2(A; \mu)$, respectively $\hat{q} : \hat{\Gamma}^2(A; \hat{\mu}) \to \widetilde{\hat{\Gamma}^2}(A; \hat{\mu})$ be the canonical unitary linear mappings onto the corresponding Hilbert direct integrals of the fields $\Gamma^2(A; \mu)$, resp., $\hat{\Gamma}^2(A; \hat{\mu})$; here $\widetilde{\Gamma}^2(A; \mu)$, resp., $\widetilde{\hat{\Gamma}^2}(A; \hat{\mu})$ are the associated (Hausdorff, complete) Hilbert spaces and they are, sometimes, less properly denoted as

$$\int_{E_0(A)}^{\oplus} H_f \, d\mu(f), \quad \text{resp.,} \quad \int_{F(A)}^{\oplus} H_f \, d\hat{\mu}(f).$$

It is easy to see that the mappings

$$u : \Gamma_0(A) \ni \theta(a) \mapsto \pi_{f_0}(a)\xi_{f_0}^0 \in H_{f_0} \quad a \in A,$$

and

$$\hat{u} : \hat{\Gamma}_0(A) \ni \hat{\theta}(a) \mapsto \pi_{f_0}(a)\xi_{f_0}^0 \in H_{f_0}, \quad a \in A,$$

are correctly defined unitary linear mappings. They extend in a unique manner to unitary linear mappings

$$U_\mu : \Gamma^2(A; \mu) \to H_{f_0} \text{ and } \hat{U}_{\hat{\mu}} : \hat{\Gamma}^2(A; \hat{\mu}) \to H_{f_0},$$

which clearly factorize through q, respectively $\hat{q}$

$$U_\mu = \widetilde{U}_\mu \circ q, \qquad \text{resp. } \hat{U}_{\hat{\mu}} = \widetilde{\hat{U}}_{\hat{\mu}} \circ \hat{q},$$

where $\widetilde{U}_\mu : \widetilde{\Gamma}^2(A; \mu) \to H_{f_0}$, resp., $\widetilde{\hat{U}}_{\hat{\mu}} : \widetilde{\hat{\Gamma}}^2(A; \hat{\mu}) \to H_{f_0}$ are uniquely determined unitary isomorphisms of Hilbert spaces.

172

The preceding construction yields the *central reduction* of the representation π_{f_0}: namely, we can consider, for any $a \in A$ the field of operators $(\pi_f(a))_{f \in F(A)}$ which is $\hat{\mu}$-*integrable* in the sense that

$$(\xi_f)_{f \in F(A)} \in \hat{\Gamma}^2(A; \hat{\mu}) \Rightarrow (\pi_f(a)\xi_f)_{f \in F(A)} \in \hat{\Gamma}^2(A; \hat{\mu}),$$

for any $a \in A$. It is easy to see that we have

(5)
$$\hat{U}\left[(\pi_f(a)\xi_f)_{f \in F(A)}\right] = \pi_{f_0}(a)\hat{U}\left[(\xi_f)_{f \in F(A)}\right],$$

for any $a \in A$ and any $(\xi_f)_{f \in F(A)} \in \hat{\Gamma}^2(A; \hat{\mu})$.

By Theorem 6, with the help of the considerations made in §4.5, we infer that there exists a surjective homomorphism of von Neumann algebras $\hat{\Phi} : Z(A^{**}) \to L^\infty(\hat{\mu})$ such that

(6)
$$\hat{U}\left[(\pi_f(a)\hat{\Phi}(z)(f)\xi_f)_{f \in F(A)}\right] = \pi_{f_0}''(z)\pi_{f_0}(a)\hat{U}\left[(\xi_f)_{f \in F(A)}\right],$$

for any $a \in A, z \in Z(A^{**})$, and $(\xi_f)_{f \in F(A)} \in \hat{\Gamma}^2(A; \hat{\mu})$.

(Here the function $F(A) \ni f \mapsto \hat{\Phi}(z)(f)$ is, for any $z \in Z(A^{**})$, an arbitrarily chosen $\hat{\mu}$-measurable representative of $\hat{\Phi}(z)$).

Formula (5) yields the central reduction (disintegration) of the representation π_{f_0}. Formula (6) shows that by this disintegration *the von Neumann algebra of the diagonalizable operators corresponds to the center of $\pi_{f_0}(A)''$.*

Given a field of operators $\hat{a} = (a_f)_{f \in F(A)}, a_f \in \mathcal{L}(H_f), f \in F(A)$, we shall say that it is $\hat{\mu}$-*integrable* (here $\hat{\mu}$ is the measure on $F(A)$ corresponding to a central measure $\mu \in \mathcal{M}_+^1(E_0(A))$, such that $f_0 = b(\mu) \in E(A)$), if the following conditions hold

(a) $\hat{\mu}$-vrai $\sup\{\|a_f\|; f \in F(A)\} < +\infty$; and
(b) $(\xi_f)_{f \in F(A)} \in \hat{\Gamma}^2(A; \hat{\mu}) \Rightarrow (a_f\xi_f)_f \in \hat{\Gamma}^2(A; \hat{\mu})$.

It is obvious that to any such field of operators there corresponds an operator $\widetilde{\hat{a}}_\mu \in \mathcal{L}(\widetilde{\hat{\Gamma}^2}(A; \hat{\mu}))$. It is easy to see that $\widetilde{\hat{U}}\widetilde{\hat{a}}_\mu\widetilde{\hat{U}}^{-1} \in (\pi_{f_0}(A)' \cap \pi_{f_0}(A)'')'$.

The field of operators $(a_f)_{f \in F(A)}$ will be said to be *universally centrally integrable*, if it is $\hat{\mu}$-integrable, for any central measure $\mu \in \mathcal{M}_+^1(E_0(A))$ such that $b(\mu) \in E(A)$.

Remark. In the preceding argumentation we had to consider also the integrable field of Hilbert spaces (0) in order to derive with the help of Theorem 4.1 from [28] the implication (4) which means that $\hat{\Gamma}(A; \hat{\mu})$ is an $\mathcal{L}^\infty(\hat{\mu})$-module. This property, together with the completeness with respect to the semi-norm $(2'')$ is an essential ingredient in Wils' definition of the integrable fields of Hilbert spaces (see [39]; and, also [28], §4).

III. As one can easily see, the central reduction of a representation π_{f_0} of A can be carried out also with help of the measure $\hat{\mu}_0 : \hat{\mathcal{B}}_0(F^0(A)) \to [0,1]$. The need for an extension of $\hat{\mu}_0$ appears when one wants to disintegrate extensions of π_{f_0} to larger C^*-algebras. We shall prove now that with the help of the completion of the measure $\hat{\mu} : \mathcal{B}(F(A)); \hat{Z}(F(A))) \to [0,1]$ one can carry out the central disintegration of the representation $\pi_f'' : \mathcal{B}(A) \to \mathcal{L}(H_{f_0})$; i.e., of the restriction of the representation π_{f_0}'' to the Borel enveloping C^*-algebra $\mathcal{B}(A)$ of A. We refer to [36] for the results on $\mathcal{B}(A)$ we shall use below.

Lemma 20. *Let K be any compact convex subset of a Hausdorff locally convex topological real vector space. Let $\mu \in \mathcal{M}_+^1(K)$ be any (Choquet–Meyer) maximal Radon probability measure and let $h_0 : K \to \mathbf{R}$ be any semi-continuous affine real function. Then h_0 is measurable with respect to the completion $\mathcal{B}_1(K)(\mu)$ of $\mathcal{B}_1(K)$ with respect to $\mu|\mathcal{B}_1(K)$.*

Proof: We recall that $\mathcal{B}_1(K)$ is the σ-algebra of subsets of K generated by $\mathcal{B}_0(K)$ and by the set $\mathcal{F}(K)$ of all compact extremal subsets of K. For the proof we shall use the notations and results from ([31], p. 11-14). Indeed, let $\mu_1 = \mu|\mathcal{B}_1(K)$. Then by ([31], Theorem 2), we have that $(\mu_1)_*(F_a) = \mu(F_a), a \in \mathbf{R}$.

In order to prove that $\mu_1^*(F_a) \leq \mu(F_a)$, it will be sufficient (equivalent) to prove that $\mu(E_a) \leq (\mu_1)_*(E_a)$. To this end we shall consider the set $G_a' = G_a \cap \Gamma(h_0)$. It is obvious that $E_a = p(G_a')$. Since G_a is a compact Baire measurable subset of K_0, by ([29], Corollary to Theorem 1), for any $\epsilon > 0$, there exists a compact extremal subset $D_1' \subset G_a$, such that $\nu(D_1) > \nu(G_a) - \epsilon$. Then, for $D_0' = D \cap D_1'$ we have that D_0' is a compact extremal subset of $G_a \cap \Gamma(h_0)$ and $\nu(D_0') > \nu(G_a) - 2\epsilon$. We infer that $p(D_0')$ is a compact extremal subset of E_a, and $\mu(p(D_0')) > \mu(E_a) - 2\epsilon$. We immediately infer that $(\mu_1)_*(E_a) \geq \mu(E_a)$ and the Lemma is proved.

Remark 1. As proved in ([19], Satz 2.1), any semi-continuous affine real function $h_0 : K \to \mathbf{R}$ is bounded. This is an immediate consequence of the barycentric calculus, which holds for such functions.

Remark 2. The preceding lemma was contained in an answer given to a question put by G. Pǎltineanu in a private conversation.

We recall that by A^m_{sa} one denotes the subset of all lower semi-continuous elements in A^{**}_{sa}, over A; by $\mathcal{U}(A)$ one denotes the real vector subspace of A^{**}_{sa}, consisting of all the (strongly) universally measurable elements over A (see [22], p. 104; [36], p. 7).

Lemma 21.

(a) For any $a \in \mathcal{U}(A)$, the function $\lambda_A(a)$ is $(\mu|\mathcal{B}_2(E_0(A)))$-measurable for any central measure $\mu \in \mathcal{M}^1_+(E_0(A))$ such that $f = b(\mu) \in E(A)$.

(b) For any $a \in \mathcal{U}(A)$ the function $\lambda_A(a)|F(A)$ is $\hat{\mu}$-measurable for any central measure $\mu \in \mathcal{M}^1_+(E_0(A))$ such that $f = b(\mu) \in E(A)$.

(c) We have that

(∗)
$$f(a) = \int_{E_0(A)} \lambda_A(a)\, d\mu = \int_{F(A)} \lambda_A(a)\, d\hat{\mu},$$

for any $a \in \mathcal{U}(A)$ and any central measure $\mu \in \mathcal{M}^1_+(E_0(A))$ such that $f = b(\mu) \in E(A)$.

Proof:

(a) Let $a \in A^m_{sa}$; then there exists a (bounded) increasing net $(a_\alpha)_\alpha$ in A_{sa}, such that

$$\lambda_A(a_\alpha) \uparrow \lambda_a(a)$$

on $E_0(A)$. Let $b_\alpha = \pi_u(a_\alpha) \in B$; then $(b_\alpha)_\alpha$ is a (bounded) increasing net in B and, therefore, we have $b_\alpha \uparrow b$ in B^{**}_{sa}, where $b \in B^m_{sa}$. It is easy to see that

(1)
$$\lambda_A(a) \circ s = \lambda_B(b).$$

Let $\nu \in \mathcal{M}^1_+(E(B))$ be the maximal orthogonal measure on $E(B)$ corresponding to μ, as in §6.4. Then we have $s_*(\nu) = \mu$ and by Lemma 20, the function $\lambda_B(b)$ is $(\nu|\mathcal{B}_1(E(B)))$-measurable; since we have that

$$\mathcal{B}_1(E(B)) \subset \mathcal{B}_2(E(B)),$$

it follows that $\lambda_B(b)$ is $(\nu|\mathcal{B}_2(E(B)))$-measurable. By formula (1) and Theorem 29, we infer that $\lambda_A(a)$ is $(\mu|\mathcal{B}_2(E_0(A)))$-measurable.

(b) This follows immediately from statement (a) and from Theorem 26, (b).

(c) This follows immediately from the fact that the barycentric calculus holds (with respect to any measure in $\mathcal{M}^1_+(E_0(A))$) for any semi-continuous affine real function on $E_0(A)$ and also from Lemma 19.

The Lemma is thus proved for semi-continuous elements in A^{**}_{sa} (over A).

For the case of an arbitrary $a \in \mathcal{U}(A)$ one can use the method of proof given in ([31], proof of Theorem 3). The Lemma is proved.

175

Lemma 22. *For any $a \in \mathcal{U}(A)$ and any $b, c \in A$, the function*

$$F(A) \ni f \mapsto f(c^*ab) \in \mathbb{C}$$

is $\hat{\mu}$-measurable for any central measure $\mu \in \mathcal{M}^1_+(E_0(A))$, such that $f_0 = b(\mu) \in E(A)$ and we have the equality

$$f_0(c^*ab) = \int_{F(A)} f(c^*ab) \, d\hat{\mu}(f).$$

Proof: By ([36], Lemma 7), we have that

$$b^*\mathcal{U}(A)b \subset \mathcal{U}(A),$$

for any $b \in A$. From the polarization formula

$$\begin{aligned}
f(c^*ab) = \frac{1}{4}[&f((b+c)^*a(b+c)) - f((b-c)^*a(b-c)) \\
&+ if((b+ic)^*a(b+ic)) - if((b-ic)^*a(b-ic))],
\end{aligned}$$

and from Lemma 21 the assertion now immediately follows.

By analogy with the definition given in ([36], p. 24), we shall say that an element $a \in A^{**}$ is *universally centrally disintegrable* if the following conditions are satisfied

(a) The field of operators $(\pi_f''(a))_{f \in F(A)}$ is universally centrally integrable; and if we denote by $\pi_{\hat{\mu}_{f_0}}''(a)$ the operator in $\mathcal{L}(\widetilde{\widehat{\Gamma}}^2(A; \hat{\mu}))$ determined by the field $(\pi_f''(a))_{f \in F(A)}$, we should also have

(b) $\widetilde{U}_{\hat{\mu}_{f_0}} \pi_{\hat{\mu}_{f_0}}''(a) = \pi_{f_0}''(a) \widetilde{U}_{\hat{\mu}_{f_0}}$, where $f_0 = b(\mu_{f_0})$ is the barycenter of the central measure $\mu_{f_0} \in \mathcal{M}^1_+(E(A))$ to which $\hat{\mu}_{f_0}$ corresponds, for any $f_0 \in E(A)$.

With a similar proof as for ([36], Lemma 5), we can state

Lemma 23.

(i) *If $a, b \in A^{**}$ are universally centrally disintegrable, then $a + b$ and ab are universally centrally disintegrable, and αa is universally centrally disintegrable for any $\alpha \in \mathbb{C}$.*

(ii) *The norm limit of a sequence of universally centrally disintegrable elements of A^{**} is universally centrally disintegrable.*

(iii) *If $(a_n)_{n \in \mathbb{N}}$ is a bounded monotone sequence of universally centrally disintegrable elements in A^{**}_{sa}, then its limit is a universally centrally disintegrable element.*

We can also prove

Lemma 24. *For any universally centrally disintegrable element $a \in A^{**}$ the formula*

$$\int_{F(A)} f(c^*ab) \, d\hat{\mu}_{f_0}(f) = f_0(c^*ab)$$

holds for any $b, c \in A$. In particular, the "central barycentric calculus" holds for a; i.e.,

$$f_0(a) = \int_{F(A)} f(a) \, d\hat{\mu}_{f_0}(f).$$

Proof: Since $(\theta_f(b))_{f \in F(A)} \in \hat{\Gamma}^2(A; \hat{\mu}_{f_0})$, we also have $(\pi_f''(a)\theta_f(b))_{f \in F(A)} \in \hat{\Gamma}^2(A; \hat{\mu}_{f_0})$ and for any $b, c \in A$ we shall have

$$\int_{F(A)} f(c^*ab) \, d\hat{\mu}(f) = \int_{F(A)} (\pi_f''(a)\theta_f(b)|\theta_f(b))_f \, d\hat{\mu}(f)$$
$$= ((\pi_f''(a)\theta_f(b))_f | (\theta_f(c))_f) = (\hat{U}_{\hat{\mu}}(\pi_f''(a)\theta_f(b))_f | (\theta_f(c))_f)$$
$$= (\pi_{f_0}''(a)\theta_{f_0}(b)|\theta_{f_0}(c)) = f_0(c^*ab).$$

On the other hand, since we have that $(\xi_f^0)_{f \in F(A)} \in \hat{\Gamma}^2(A; \hat{\mu})$, (see [28], Proposition 4.5 and Theorem 4.3), we immediately infer that

$$\int_{F(A)} f(a) \, d\hat{\mu}_{f_0}(f) = f_0(a).$$

The Lemma is proved.

Let $\mathcal{C}(A) \subset A^{**}$ be the set of all universally centrally disintegrable elements in A^{**}. We define

$$\mathcal{C}_0(A) = \mathcal{C}(A) \cap \mathcal{C}(A)^*,$$

where $\mathcal{C}(A)^* = \{a^* \in A^{**}; a \in \mathcal{C}(A)\}$.

Theorem 32. $\mathcal{C}_0(A)$ *is a C^*-algebra whose self-adjoint part is sequentially monotone closed.*

Proof: Immediate consequence of Lemma 23.

As in ([36], p. 7) we denote by $\mathcal{B}_{sa}^0(A)$ the smallest real vector subspace of A_{sa}^{**} which contains $(A_{sa})^m$ and is closed with respect to the sequential bounded monotone convergence in A_{sa}^{**}; convergence which is to be understood either with respect to the order relation in A_{sa}^{**} or, equivalently, strongly on the space H_u.

Lemma 25. *Any $a \in \mathcal{B}_{sa}^0(A)$ is universally centrally disintegrable.*

Proof: By ([36], §4) we have that

$$\mathcal{B}_{sa}^0(A) \subset \mathcal{U}(A)$$

and, by ([36], Proposition 1) we have that

$$a \in \mathcal{B}_{sa}^0(A) \Rightarrow a^2 \in \mathcal{B}_{sa}^0(A).$$

We shall now consider the vector space $\hat{\Gamma}_a(A) \subset \Pi_{f \in F(A)} H_f$ defined by

$$\hat{\Gamma}_a(A) = \{(\pi_f''(a)\theta_a(b) + \theta_f(c))_{f \in F(A)}; b, c \in A\}.$$

It is obvious that

$$\hat{\Gamma}_0(A) \subset \hat{\Gamma}_a(A) \subset \Pi_{f \in F(A)} H_f.$$

Since we have that

$$f((ab_2 + c_2)^*(ab_1 + c_1)) = f(b_2^* a^2 b_1) + f(b_2^* ac_1) + f(c_2^* ab_1) + f(c_2^* c_1),$$

for any $f \in E_0(A)$ and any $b_1, b_2, c_1, c_2 \in A$ we infer that the function

$$F(A) \ni f \mapsto (\xi_f | \eta_f)_f \in \mathbb{C}$$

is $\hat{\mu}$-integrable, for any $(\xi_f)_{f \in F(A)}, (\eta_f)_{f \in F(A)} \in \hat{\Gamma}_a(A)$. We can, therefore, consider the L^2-completion $\hat{\Gamma}_a^2(A; \hat{\mu})$ of $\hat{\Gamma}_a(A)$ with respect to $\hat{\mu}$ (see [28], §4). We can define correctly a linear mapping

$$V_a : \hat{\Gamma}_a(A) \to H_{f_0},$$

by the formula

$$V_a[(\pi_f''(a)\theta_f(b) + \theta_f(c))_{f \in F(A)}] = \pi_{f_0}''(a)\theta_{f_0}(b) + \theta_{f_0}(c),$$

for any $b, c \in A$. The correctness of the definition follows from the fact that if $f((ab + c)^*(ab + c)) = 0$, for any $f \in F(A)$, then, by Lemma 24 we have

$$f_0((ab + c)^*(ab + c)) = \int_{F(A)} f((ab + c)^*(ab + c)) \, d\hat{\mu}_{f_0}(f) = 0,$$

and this implies that $\pi_{f_0}^{**}(a)\theta_{f_0}(b) + \theta_{f_0}(c) = 0$.

178

It is easy to see that V_a is an isometric lienar mapping; therefore, it extends uniquely to an isometric linear surjective mapping

$$W_a : \hat{\Gamma}_a^2(A; \hat{\mu}) \to H_{f_0}.$$

We infer that

$$\hat{\Gamma}^2(A; \hat{\mu}) = \hat{\Gamma}_a^2(A; \hat{\mu}),$$

and, therefore, for any $b \in A$ there exists a strongly integrable vector field $(\xi_f)_{f \in F(A)} \in \hat{\Gamma}^2(A; \hat{\mu})$ such that $\pi_f''(a)\theta_f(b) = \xi_f, \hat{\mu}$, a.e. We infer that

$$(\xi_f)_{f \in F(A)} \in \hat{\Gamma}^2(A; \hat{\mu}) \Rightarrow \left(\pi_f''(a)\xi_f \right)_{f \in F(A)} \in \hat{\Gamma}^2(A; \hat{\mu}).$$

On the other hand, from the formula

$$\int_{F(A)} f(b^* a^2 b) \, d\hat{\mu}_{f_0}(f) = f_0(b^* a^2 b) = \|\pi_{f_0}''(a)\theta_{f_0}(b)\|^2,$$

which holds for any $b \in A$, we infer that

$$\widetilde{\hat{U}}_{\hat{\mu}_{f_0}} \pi_{\hat{\mu}_{f_0}}''(a) = \pi_{f_0}''(a) \widetilde{\hat{U}}_{\hat{\mu}_{f_0}},$$

and the Lemma is proved.

We can now prove the main result of the paper:

Theorem 33. *Any $a \in \mathcal{B}(A)$ is universally centrally disintegrable.*

Proof: Let us define

$$\mathcal{A}(A) = \left\{ \sum_{k=1}^{n_1} a_{k1}' \ldots a_{km_1}' + i \sum_{k=1}^{n_2} a_{k1}'' \ldots a_{km_2}''; a_{kj}', a_{kj}'' \in \mathcal{B}_{sa}^0(A) \right\};$$

then $\mathcal{A}(A)$ is a $*$-subalgebra of $\mathcal{B}(A)$ and its norm closure $\mathcal{B}_1(A)$ is a C^*-subalgebra of $\mathcal{B}(A)$. From Lemma 25 and from part (i) of Lemma 23 we infer that any element $a \in \mathcal{A}(A)$ is universally centrally disintegrable. From part (ii) of Lemma 25 we infer that any element in $\mathcal{B}_1(A)$ is universally centrally disintegrable. If we denote

$$\mathcal{M} = \{a \in \mathcal{B}(A)_{sa}; a \text{ is univ. centr. disintegr.}\},$$

we infer that

$$\mathcal{B}_1(A)_{sa} \subset \mathcal{M} \subset \mathcal{B}(A)_{sa}.$$

By part (iii) of Lemma 25 we infer that $\mathcal{M}$ is a sequentially montone closed subset of A_{sa}^{**}. Lemma 4 from [36] implies that $\mathcal{M} = \mathcal{B}(A)_{sa}$, whereas part (i) of Lemma 25 now ends the proof.

Remark. The preceding theorem is a generalization to the general, possibly non-separable, case of a theorem of Sakai (see [24], Theorem 3.5.2). It is clear that even for the commutative (non-separable) case the use of the measure $\hat{\mu}$, instead of the measure $\hat{\mu}_0$ is essential for the obtention of this generalization.

IV. For the elements belonging to the center $Z(\mathcal{B}(A))$ of the C^*-algebra $\mathcal{B}(A)$ we can obtain an explicit expression for the element $\Phi_{f_0}(a) \in L^\infty(\hat{\mu}_{f_0})$, given by Sakai's mapping $\Phi_{f_0}, f_0 \in E(A)$ (see §2.4, above).

Indeed, let us first remark that for any $z \in Z(A^{**})$ and any $a \in A^{**}$, we have that

$$f(za) = f(z)f(a), \qquad f \in F(A).$$

Let now $f_0 \in E(A)$ and $z \in Z(\mathcal{B}(A))$ be given. Then, by Lemma 24 and Theorem 33 we have that

$$
\begin{aligned}
f_0(za) &= \int_{E_0(A)} \lambda_A(za) \, d\mu_{f_0} = \int_{F(A)} \lambda_A(za) \, d\hat{\mu}_{f_0} \\
&= \int_{F(A)} \lambda_A(z)\lambda_A(a) \, d\hat{\mu}_{f_0} = \left(K_{\mu_{f_0}}([\lambda_A(z)]) \, \pi_{f_0}(a)\xi_{f_0}^0 | \xi_{f_0}^0\right) \\
&= \left(\pi_{f_0}''(z)\pi_{f_0}(a)\xi_{f_0}^0 | \xi_{f_0}^0\right),
\end{aligned}
$$

for any $a \in A$ and this implies that

$$\Phi_{f_0}(z) = [\lambda_A(z)] \qquad (\mathrm{mod}\ \hat{\mu}_{f_0}).$$

We have thus proved the following

Theorem 34. *For any state $f_0 \in E(A)$ and any $z \in Z(\mathcal{B}(A))$ we have that*

$$\Phi_{f_0}(z) = [\lambda_A(z)] \qquad (\mathrm{mod}\ \hat{\mu}_{f_0})$$

on $F(A)$.

Remark. It would be interesting to see whether the equality

$$\Phi_{f_0}(Z(\mathcal{B}(A))) = L^\infty(\hat{\mu}_{f_0})$$

holds for any $f_0 \in E(A)$. An affirmative answer would give an improved version to the description of Sakai's mapping Φ_{f_0}.

References

1. E. M. Alfsen, *Compact Convex Sets and Boundary Integrals.* Springer Verlag, Berlin–Heidelberg–New York, 1971 (Ergebnisse der Mathematik un ihrer Grenzgebiete, Bd. 57).

2. J. Anderson and J. W. Bunce, Stone–Weierstrass theorems for separable C^*-algebras, Journ. Oper. Theory **6** (1981), 363–374.

3. R. J. Archbold, C. J. K. Batty, On factorial states of operator algebras, III, preprint, 1984, 0–43).

4. L. Asimow and A. J. Ellis, Convexity Theory and its Applications in Functional Analysis, Academic Press, London, 1980.

5. C. J. K. Batty, Some properties of maximal meaures on compact convex sets, Math. Proc. Cambridge Phil. Soc. **94** (1983), 297–305.

6. C. J. K. Batty, Topologies and continuous functions on extreme points and pure states, preprint, 1984, 0–25.

7. C. J. K. Batty and R. J. Archbold, On factorial states of operator algebras, II., Journ. Oper. Theory **13** (1985), 131–142.

8. N. Boboc and Gh. Bucur, Cônes convexes de fonctions continues sur un espace compact, topologies sur la frontière de Choquet, Rev. Roumaine Math. Pures Appliquées **17** (1972), 1307–1316.

9. N. Boboc and Gh. Bucur, Conuri convexe de functii continue pe spatii compacte, Ed. Acad. RSR, Bucharest, 1976.

10. O. Bratteli and D. W. Robinson, *Operator Algebras and Quantum Statistical Mechanics* I., Springer Verlag, Berlin–Heidelberg–New York, 1979.

11. J. Dixmier, Les algèbres d'opérateurs dans l'espace Hilbertien, Gauthier–Villars, Paris, 2^{ieme} éd., 1969.

12. J. Dixmier, Les C^*-algèbres et leurs représentations, Gauthier–Villars, Paris, 2^{ieme} éd., 1969.

13. E. G. Effros, Structure in simplexes, Acta Math. **117** (1967), 103–121.

14. E. G. Effros, Structure in simplexes II, Journ. Funct. Analysis **1** (1967), 361–391.

15. E. G. Effors and A. Gleit, Structure in simplexes, Trans. Amer. Math. Soc. **142** (1969), 355-379.

16. A. Gleit, On the structure topology of simplex spaces, Pacific Journ. Math. **34** (1970), 389–405.

17. R. Godement, Les fonctions de type positif et la théorie des groupes, Trans. Amer. Math. Soc. **63** (1948), 1–84.

18. R. W. Henrichs, On decomposition theory for unitary representations of locally compact groups, Journ. Funct. Analysis **31** (1979), 101–114.

19. U. Krause, Der Satz von Choquet als ein abstrakter Spektralsatz und vice versa, Math. Annalen **4** (1970), 275–296.

20. P. A. Meyer, *Probability and Potentials.* Blaisdell Publ. Comp., Waltham–Toronto–London, 1966.

21. G. Paltineanu, Elemente de teoria approximarii functiilor continue, Ed. Acad. RSR, Bucharest, 1982.

22. G. K. Pedersen, *C*-algebras and Their Automorphism Groups.* Academic Press, London–New York–San Francisco, 1979.

23. R. R. Phelps, *Lectures on Choquet's Theorem.* D. van Nostrand Co., Princeton–Toronto–New York–London, 1966.

24. S. Sakai, *C*-algebras and W*-algebras.* Springer Verlag, Berlin–Heidelberg–New York, 1971.

25. C. F. Skau, Orthogonal measures on the state space of a *C**-algebra, *Algebras in Analysis*, ed. by W. Williamson, Academic Press, London–New York–San Francisco, 1975.

26. M. Talagrand, Pettis integral and measure theory, Mem. Amer. Math. Soc. **51** (1984).

27. M. Takesaki, *Theory of Operator Algebras* I, Springer Verlag, New York–Heidelberg–Berlin, 1979.

28. S. Teleman, An introduction to Choquet theory with applications to Reduction Theory, INCREST Preprint Series in Mathematics **71** (1980), 1–294.

29. S. Teleman, On the regularity of the boundary measures. INCREST Preprint Series in Mathematics **30** (1981), 1–22. Appeared in "Lecture Notes in Mathematics **1014**, Springer Verlag, Berlin–Heidelberg–New York–Tokyo (1983), 296–315.

30. S. Teleman, Measure-theoretic properties of the Choquet and of the maximal topologies. INCREST Preprint Series in Mathematics **33** (1982), 1–41.

31. S. Teleman, On the non-commutative extension of the theory of Radon measures. INCREST Preprint Series in Mathematics **1** (1983), 1–21.

32. S. Teleman, A lattice-theoretic characterization of Choquet simplexes. INCREST Preprint Series in Mathematics **37** (1983), 1–15.

33. S. Teleman, Measure-theoretic properties of the maximal orthogonal topology. IN-CREST Preprint Series in Mathematics **2** (1983), 1–35.

34. S. Teleman, On the Choquet and Bishop–de Leeuw Theorems. In the volume "Spectral Theory", Banach Center Publications **8** (1982), PWN–Polish Scientific Publishers, Warsaw, 455–466.

35. S. Teleman, Topological properties of the boundary measures. In the volume "Studies in probability and related topics". Papers in Honour of Octav Onicescu on his 90th birthday, Nagard Publisher, 1983, 457–463.

36. S. Teleman, On the Borel enveloping C^*-algebra of a C^*-algebra. INCREST Preprint Series in Mathematics **35** (1984), 1–49.

37. S. Teleman, On the Dauns–Hofmann Theorem. INCREST Preprint Series in Mathematics **41** (1984), 1–9.

38. W. Wils, Désintégration centrale des formes positives sur les C^*-algebras, C. R. Acad. Sci. Paris **267** (1968), 810–812.

39. W. Wils, Direct integrals of Hilbert spaces, I. Math. Scand. **26** (1970), 73–88.

40. W. Wils, Désintégration centrale dans une partie convexe compacte d'un espace localement convexe, C. R. Acad. Sci. Paris **269** (1969) 702–704.

41. W. Wils, The ideal center of partially ordered vector spaces, Acta Mathem. **127** (1971), 41–77.

42. L. Zsido, Descompuneri topologice ale W^*-algebrelor I. Studii si Cercetari matem. **25** (1973), 859–945; idem II. Studii si Cercetari matematice **25** (1973), 1037–1112.

Silviu Teleman
Department of Mathematics
University of Puerto Rico
P. O. Box 23355
Rio Piedras, San Juan, PR 00931